# Intelligent Systems for Crisis Management

# Intelligent Systems for Crisis Management

Editors

**Raj Patil and Manzar Khan**

# Intelligent Systems for Crisis Management

Edited by **Raj Patil and Manzar Khan**

Printed in 2017

ISBN: 978-1-68117-130-2

Library of Congress Control Number: 2015936551

© 2016 by

SCITUS Academics LLC,
616, Corporate Way, Suite 2, 4766,
Valley Cottage, NY 10989

www.scitusacademics.com

**Notice**

# Contents

vi

# Preface

Many aspects related to the efficient collection and integration of geo-information, applied semantics and situation awareness for disaster management are still open. To advance the systems and make them intelligent, an extensive collaboration is required between emergency responders, disaster managers, system designers and researchers. To facilitate this process the Geo-information for Disaster Management (Gi4DM) conference has been organized since 2005. Gi4DM is coordinated by the Joint Board of Geospatial Information Societies (JB GIS) and the ad-hoc Committee on Risk and Disaster Management. This volume presents the results of the Gi4DM 2012 conference, held in Enschede, the Netherlands, on 13-15 December. It contains a selection of around 30 scientific and 25 best-practice peer-reviewed papers. The 2012 Gi4DM focuses on the intelligent use of geo-information, semantics and situation awareness.

**Editor**

# Chapter 1

# Risks and Crises for Healthcare Providers: The Impact of Cloud Computing

Ronald Glasberg, Michael Hartmann, Michael Draheim, Gerrit Tamm, and Franz Hessel

SRH Hochschule Berlin, 10587 Berlin, Germany

## ABSTRACT

We analyze risks and crises for healthcare providers and discuss the impact of cloud computing in such scenarios. The analysis is conducted in a holistic way, taking into account organizational and human aspects, clinical, IT-related, and utilities-related risks as well as incorporating the view of the overall risk management.

## INTRODUCTION

In the industrialized countries hospitals are the backbone of the healthcare system. In Germany 18.620.422 hospital treatments were conducted in 2.017 hospitals in 2012 [1]. Like in most countries, nearly half of the hospital beds are in public ownership with a growing number of privately owned hospitals [2]. The aim of the hospitals is to heal diseases, prevent their deterioration, or alleviate disease symptoms, with specialized staff and equipment. For that reason, hospitals are a relatively hazardous working environment for patients as well as staff. The hospital staff has to deal with adverse events and numerous potential, for example, wound infections, medication errors, and wrong-site surgery [3, 4]. This permanent risk of unsafe situations makes the hospital sector an important setting for an assessment for safety and risk management. The majority of the publications and studies on risk management in hospitals addressed clinical safety and risk management in specific indications, medical subspecialties, or treatment settings such as intensive care or operation theatre [5–7]. Despite this substantial body of research in the area of patient safety in specific situations there are only a small number of systematic reviews or comprehensive, interdisciplinary approaches. Based on a systematic literature review Hoff and colleagues postulated that primarily the interventions and not the organizational structure and features are linked directly to patient safety [8]. In a more recent work, Dückers and colleagues draw the somehow frustrating conclusion that the scientific evidence for safety interventions in hospitals still is limited and that the methodological quality of the studies is generally weak [9]. Although a recent hospital survey indicates increased attention to the management of risks in hospitals, we are far from having defined a general approach for various sources of risks, their analysis, evaluation, and treatment [10].

Not only the risks directly related to patient treatment, but also the continuous governmental healthcare reform acts and the increasing financial pressure on healthcare spending are big challenges for a sustainable hospital management. On the other side, information technology innovations are often considered as a major factor for the improvement of quality, efficiency, and efficacy in healthcare [11]. As one approach electronic health records (EMR) promise to improve efficiency and effectiveness of healthcare providing processes [12]. The use of electronic data in hospitals is ubiquitous and inevitable and the use of health IT is still increasing but according to the most recent data still only less than one-third of the hospitals in the US use a kind of electronic medical records [13]. Due to the slow speed of implementation of information technology the expected massive cost savings by EMR did not yet come true [14]. With regard to the quality of patient care there is only marginal improvement, too [15]. In particular, medical doctors seem to be relatively reluctant to leave the traditional way of unstructured paper-pencil documentation [16] and to adopt IT technologies in daily patient care. Other approaches go as far as incorporating virtual or mixed reality [17] as well as intelligent systems [18] in healthcare scenarios. Cloud computing (CC) is increasingly being viewed as a key innovation in this regard and is generally considered one of the most important developments in IT [19]. But in addition to the opportunities that information technology pervasion to a hospital, these new technologies also pose risks to the organisations. Security and privacy are the relevant threats for hospitals in such a cloud environment, because health data are the most private and sensitive data about the patients [20].

In our project, we extended the scope of the potential use of health IT and cloud computing in hospitals, from the "classical" objectives of cost savings, quality management, and clinical risk management to hospital crises. Objectives of the project are to identify the specific crisis scenarios perceived as most relevant by hospital care providers, to evaluate the preparedness of hospitals to prevent, respectively, handle the crisis scenarios, and to describe and develop IT and cloud computing solutions to support crisis management in hospitals. The specific focus of this paper lies on identification and handling of IT crises.

In general, a crisis is described as "an abnormal situation, or even perception, which is beyond the scope of everyday business and which threatens the operation, safety, and reputation of an organisation" [21]. Transferred to hospital management, a crisis is one or numerous critical situations which could not be handled by routine measures of quality management. A hospital crisis is regarded as an event or a series of events, which may occur either suddenly or which may take some time to evolve. It results in a major, urgent problem with potentially severe consequences for the hospital and it must be addressed immediately.

Hospital crises can roughly be categorized into natural disasters (i.e., earthquakes, floods, or fires), significant operational problems (i.e., personnel emergencies, accidents, and theft of proprietary data) or extraordinary problems (i.e., hostage situations) [22]. To identify all relevant crises to a hospital, it is also necessary to address internal problems. As visualized in Figure 1, we classified hospital crises into four areas according to the professional disciplines affected by the crisis: medical care, information systems (IS), human resources (HR), and supply. In a contribution with the use of CC in hospitals we present our results from the area of Information Systems and Supply.

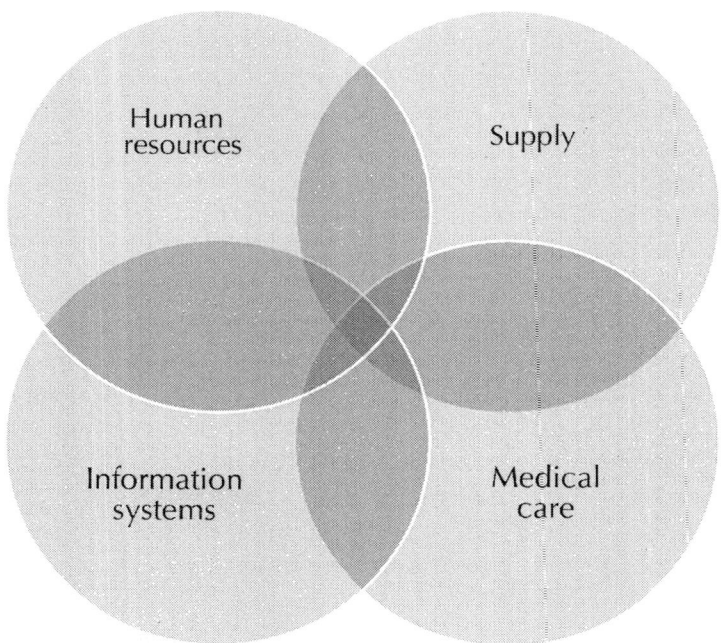

**Figure 1**: Considered areas in a hospital.

The rest of this work is structured as follows. In Section 2 risks associated with IT and utilities are discussed. Section 3 gives an overview of the overall approach and in Section 4 the results of the evaluation are presented. Finally in Section 5 we discuss our results and future research activities.

Risks and Crises for Healthcare Providers: The Impact of Cloud...

5

# IT-RELATEDANDUTILITY-RELATEDRISKS

When running IT systems which process health data, both the original organization (e.g., a clinic) and the CC provider should implement a number of appropriate technical and organizational measures of precaution. For data originating from German healthcare applications these measures are specified in a catalog of eight control requirements (§ 9 in conjunction with the annex to § 9 BDSG—the German Data Protection Law). There is a similar requirement for socially related data in § 78a of the Social Codex (Sozialgesetzbuch, SGB), in conjunction with the annex to § 78a SGB.

Associated risks in this context arise from the fact that the law stipulates only general requirements. The precise definition and implementation of specific measures are the obligation of both the healthcare provider and the CC provider. For example, both should apply general measures for protecting personal data (e.g., limited access) when dealing with health-related data and also implement measures to protect data transmission (e.g., encryption). Furthermore, systems that are operated for more than one client (e.g., processing appointment data or analysis data for multiple clinics) should ensure strict separation between data of each client organization. There exist specific recommendations about the compliant operation of a hospital management system (HIS) [23]. Similar requirements apply for CC and outsourcing scenarios, as providers are expected to implement and assure security requirements of the client organization.

Specific risks and crises can occur when the organization is not capable to follow all applicable regulations in the area of medical confidentiality, social data, and state-specific rules (rules that are different in every specific German federal state).

Medical confidentiality describes the relation of trust between a doctor and a patient. In Germany it is regulated in the professional code of conduct for doctors (Muster-Berufsordnung für die deutschen Ärzte und Ärztinnen (MBO-Ä)) with medical confidentiality specified in § 9 Section.1 MBO-Ä. A breach of confidentiality is considered a criminal offense and a reveal of patient data can result already from archiving patient data with a service provider without the previous written consent of the patient. This previous written consent should include the specific data and the legal information about the service provider and is therefore often unfeasible.

Social data as a term includes all personally related data that concerns social aspect of a person. The increased confidentiality requirements for social data are defined in § 35 Sect. 1 SGB I. A specific example of regulations in this area is the recently introduced "electronic health card" (elektronische Gesundheitskarte, eGK), a reduced EMR backup on the identity card of statutory sickness fund enrollees. Requirements concerning data protection in the context of the eGK are specified in Volume V of SGB, with particular regulations concerning encryption and access control lists (ACLs) in § 291a SGB V.

Further IT-related risks and crises can also occur when a clinic neglects obligations mandated by state-specific rules (e.g., specific and different rules in Bavaria, Hamburg, or Berlin) with respect to data protection and information processing. A variety of state-specific hospital laws exist that often stipulate different requirements with respect to patient data processing. For example, according to the state hospital law (Landeskrankenhausgesetz, LKG) of Berlin, hospitals in Berlin are either allowed to process patient data in-house or outsource this process to another hospital. Other providers can process patient data under the mandate of the hospital only if it has been sufficiently anonymized in order to eliminate person-related aspects from it (§ 24 Sect. 7 (2) LKG Berlin).

Data processing in the context of CC typically constitutes the so called data processing under mandate (German: Datenverarbeitung im Auftrag) as stipulated by § 11 BDSG. This results in another wide range of risk and crisis-scenarios stemming from the specific requirements regarding the contractual relationship between client and service provider. The contract should specify the type and scope of the intended use of data, the control rights of the client, and the specific technological and organizational measures that the provider will be implementing in accordance with § 9 BDSG. Furthermore, prior to the start of the actual data processing under mandate, the client has to carefully select the service provider and to convince himself that the technological and organizational measures are appropriate (§ 11 Section 2 (4) BDSG). This control obligation continues during the actual data processing with a requirement of regularly controls. Noncompliance with it can result in regulatory fines. Major CC providers in Germany conduct yearly audits by independent audit organizations and make the audit reports available to their clients (http://www.pironet-ndh.com/site/pndh-website-site/node/269414/Lde/).

Data processing under mandate with respect to social data is regulated similarly but by § 80 SGB X. There are several important differences to § 11 BDSG that can lead to additional risk and crisis scenarios—client organizations are allowed to use in general only providers from the public administration. The inclusion of a private CC provider can only be considered if otherwise there will be substantial problems for the normal operation of the client organization or if there are substantial cost benefits in comparison to a provider from the public administration. As there are currently no reliable assessments whether private CC providers can plausibly meet these conditions, we regard their inclusion in scenarios covered by the SGB as legally unclear and thus having the potential to further amplify the impact of major IT-related risks and crises scenarios.

The presented inherent risks of cloud-based data processing for healthcare providers show that these organizations should have an elaborated demand and requirements concept with respect to data privacy. The concept should consider aspects such as the selection and evaluation process of possible CC providers, specific detailed requirements about service level agreements (SLAs), and specifically required organizational and technical measures that the CC provider should conform to. This dramatically increases transaction costs in the CC market, which is already marked by high levels of information asymmetry [24]. Some existing automated approaches for matching demand and supply, even at the level of SLAs [25], are only of limited benefit, as they cannot account properly for complex organizational measures. Specific technical measures, on the other side, can be clearly stated in automated supply statements (e.g., the so called service level objectives as introduced in [26]) and can therefore be easily matched to automated requirements. Recommendations for specific measures can be derived from relevant standards, such as the Baseline IT-Security (IT-Grundschutz) standard by the Federal Office for Information Security (Bundesamt für Informationssicherheit, BSI).

Utilities-related risks associated with operating a healthcare provider have been rarely studied, with power-supply-related incidents being considered only during intra-hospital transfer of critically ill patients

[27]. Other important utilities, for example, the supply with gases have only been considered in the context of standardization efforts for the specific case for Britain in 1979 [23], while water supply is typically assessed only as a potential source of infections (e.g., Legionellaceae) [26, 28, 29].

In our approach, we introduce the two perspectives, the IT-related and the utilities-related, into the overall model in order to better estimate the impact of crises that can arise from these fields and to better recommend appropriate countermeasures, both proactively and retroactively.

In this work we present an approach that aims to consider diverse aspects in the context of risk and crisis management for healthcare providers in the context of CC, ranging from human resources and clinical management to IT-related and utilities-related aspects. Our analysis is focused on Germany, as it is a jurisdiction with one of the most elaborated and restrictive regulations with respect to liability, data protection, and duty of care particularly in the area of healthcare [30].

# OVERVIEWOFTHEAPPROACH

The objective of our project, "Risk Management in hospitals", is to analyze the behavior of the various key players in the fields of Medical Care and Medicine, Supply, HR, and IT-Systems with regard to the influence of dynamic risks in the context of various simulated scenarios. In the first step of the research project, a network of experts and executives from politics, business, and media related to the hospital field should be established, accompanied by the creation of a literature database. In a second step, information will be collected in expert workshops and, in combination with the results of the literature search and the expert interviews, will be used for the conception, planning, and implementation of a prototype Decision-Making Tool. With such a tool (based upon artificial neural networks), safety-relevant deficits in the hospital as well as the development of ideal-solutions will be illustrated. This project will thus provide decision-making support for directing and managing bodies of hospitals. Our empirical assessment follows a qualitative approach. For the most important hospital crises, identified by literature search and interdisciplinary expert groups, we evaluate the preparedness of German hospitals and develop adequate management scenarios including IT solutions such as cloud computing. These solutions are used for on-site approaches to avoid incidents, to exchange data, as an information source, for example, for guidance documents as well as active training tools.

The core tools of our project were four expert workshops, one for each cluster of crises in IT, HR, medical care, and utilities. The participants consisted of experts and leaders of the respective fields. For the expert workshops a standardized agenda was set with the purpose of identifying the most important crises of each area. The workshops were structured in five phases, that is, the brainstorming phase, the discussion and precision phase, the evaluation phase, the dyadic phase, and the presentation phase. In the brainstorming phase the experts and managers were asked to write down all the crises they could think of. Following this, the identified crises were written on cards and clustered on pin boards by the workshop leaders. In the second phase the identified crises were presented and discussed in the whole group. The workshop leaders for the respective areas shortly presented the crisis given by the group and discussed possible ambiguities. After all, participants had the same level of knowledge about the identified crises. All of them were asked to select the most important crises from the first brainstorming in the third phase. For this purpose, each of the participants had the opportunity to award 10 points, with a maximum of 5 points for one crisis. By this vote the total number of collected crises was reduced.

In the fourth phase, the workshop participants were divided into teams of two experts. Each team of two should choose to edit two crises. The processing was done by dyadic interaction, where the two experts first worked out key features, consequences, and costs of a crisis and fixed the results on a poster. The teams had 45-minute time for the development of a single crisis. In the final workshop phase, the results of the teamwork phase were presented. For this purpose, each team introduced their findings to the group and then the results of the dyadic phase were discussed together. In the last step, the participants received the possibility to rate the danger and the probability of occurrence of the crises by setting points to a prepared evaluation scale on the posters. The final assessments served to produce a better ranking of crises.

In order to acquire participants for the workshop, a representative sample of German hospitals equally distributed with regard to ownership, number of beds, and level of care of the average population was determined. After preliminary phone calls with the managers or their assistants of hospitals (n=195) personal invitations were sent out. Overall, a number of (n=16) experts attended each of the workshops: WS 1 consisted of Medical Care and HR and WS 2 was about Information Systems and Supply. In WS 1 physicians and experts from hospital management and quality management participated. WS 2 was attended especially by heads of the IT departments as well as technical directors of hospitals.

# RESULTS

A number of specific crises in hospitals were characterizing the debates in the workshops. In particular in the area of medical malpractice, the "Use of medical devices or implants with defects or insufficient approval" and the "Occurrence of hygiene crises due to organizational deficits" were highlighted by the participants among others. All hospitals are threatened periodically by these problems which pose a significant risk to the economic survival. The fact that the participants (rather from the medical field) consider the crisis "Failure of the edp system" as one of the top-rated five crises from the field of Medical Care underlines the increasing importance of information systems in health care.

In the second expert workshop, the major crises were collected from the field of information systems and categorized according to their impact on hospitals. The results are shown in Table 1. In particular, the failure of the information technology infrastructure was identified as crisis. Furthermore, it may be discerned that the threat of cybercrime such as trojans, viruses, and also social hacking poses a relevant threat to the hospitals. Other major crisis scenarios resulting from menace arise from the treatment of patients. Also in the workshop with participants primarily from the information technology area some crises that affect the IT-support of patient treatment were identified.

**Table 1**: Results from WS2—Information systems

| Rank | Field | Risk/crisis | Description |
|------|-------|-------------|-------------|
| 1 | IS | Failure of the entire IT infrastructure, or of individual parts | The failure of the IT structures leads to the disturbance of the normal hospital workflow. The necessary flow of information is interrupted. Doctors and nurses cannot access important treatment information (such as laboratory test results). The administration cannot access rosters and accounting data. |
| 2 | IS | Trojan, virus, hacking | A criminal and defective attack on the information systems of a hospital has taken place. The data of the patient/hospital were copied, destroyed or damaged. The attacks are not or at a later time point noticed. The privacy of the patient is injured. Legal implications for the hospital might occur, if it is not proven that all necessary protective measures have been made. |

| 3 | IS | Application systems are not available | Application systems, which are necessary for adequate treatment of patients, such as the hospital information system (HIS) cannot be accessed. The technical staff of the hospital is unable to solve problems within a short time. The information (such as diagnostic images) required by the medical staff are not available. There are some limitations in the treatment as well as adequate performance documentation. |
|---|----|--------------------------------------|--------------------------------------------------------------------------------------------------------------------------------------------------------------------------------------------------------------------------------------------------------------------------------------------------------------------------------------------------------------------------------------------------------------------------------------|
| 4 | IS | Data theft/Social Hacking | Social Hacking is the acquisition of information through manipulation and deception of a person. Employees and partners have access to highly sensitive data. This approach is performed directly or through third parties. Because of carelessness or criminal activity, these data become public. For the hospital it means creating a large image damage and loss of reputation. |

| 5 | IS | Poor ergonomics lead to incorrect entries/ interpretations | Poor software ergonomics lead to incorrect entries or misinterpretation of clinical data of patients. It can increase the appearance of incorrect entries. Due to outdated systems, the probability of incorrect entries or misidentification may still increase. There will be mistreatment of patients by the poor software ergonomics. |

Another important aspect within hospital crisis management is the dependence on a variety of external resources. As shown in Chapter 1 hospitals are not only crisis-prone, they also depend on a variety of critical infrastructures. This results in a crises-evaluation in the field of supply which is shown in Table 2.

**Table 2**: Results from WS2: supply

| Rank | Field | Risk/crisis | Description |
|---|---|---|---|
| 1 | Supply | Loss of power for more than 48 hours | There is a power failure lasting longer than 48 hours. The propellant and thus the emergency power supply cannot be maintained over the entire duration of the power failure. It comes to a gradual failure of all supply elements (e.g., hot water, heating, and cooling) and communication (within and outside of the hospital). The treatment can be carried out only in a severely restricted way or not at all. |
| 2 | Supply | Heating/air falls out: evacuation necessary | Due to a failure of the heating or cooling system, an evacuation of the hospital is necessary. In consequence of a very short time frame and the threat of patient risk, an immediate action is needed. It comes to a mismatch between existing and required human resources. Scheduled treatment cannot take place and the hospital is no longer accepting patients. |

| 3 | Supply | Fire (smoke on ward) | A fire spreads out at a unit with the consequence of smoke and fire damage. Patients and staff are at risk. The unit has to be evacuated. |
|---|---|---|---|
| 4 | Supply | Failure of the water supply | In health care facilities such as hospitals, the availability of drinking water is essential to survive. A supply of water in the hospital cannot be ensured. The use of sanitary units and the execution of cleaning operations are no longer possible. While the remedy no medical processes can take place. Depending on the duration and extent of the failure, the hospital has to be evacuated. |
| 5 | Supply | Spills of dangerous substances (e.g., chlorine gas) | In many functional units of the hospital hazardous substances are used daily. These include for example disinfectants, surgical gases, drugs, and chlorine gases. There will be a release of these substances in larger quantities. The station is contaminated and needs to be evacuated. Patients and staff are directly at risk. |

The energy supply in hospitals is an element that requires a precise control, because a current reduction for some minutes or a blackout could have a significant impact due to inoperative medical equipment, hampered communications and transportation, stopped heating, and water supply. All scenarios could generate a collapse in the services. Hospitals wouldn't be able to work if they do not have a process to counter the interruption; for this reason, it is important to have a plan to mitigate and counter any emergency and also to reduce any potential risk. The "Loss of power for more than 48 hours" was highlighted by the participants as particularly important. Thus, existing fuel reserves have only to ensure the operation up to 24 hours [31]. Other key points from this workshop field were an outbreak of "fire" and the "Spills of dangerous substances". When these events occur they have a significant impact on hospitals.

# DISCUSSION AND OUTLOOK

The consideration of hospital crises in the context of cloud computing has the potential to bring new insights to decision makers in healthcare and also to enhance the body of knowledge both in the areas of healthcare management and IT management. Furthermore, by pursuing a holistic approach our work offers a framework where the implications of crises can be considered for the whole organization. The approach defines hospital crisis as an event or a series of events, which may occur either suddenly or which may take some time to evolve. It results in a major, urgent problem with potentially severe consequences for the hospital and it must be addressed immediately. The selected evaluation methodology—expert workshops—is an established approach, particularly in the area of health-related research [32]. The identified crises during the evaluation confirm expectations that problems with cloud-based systems (and IT systems in general) can lead to substantial limitations of the handling capability of a hospital. In order to further corroborate these findings the research team has launched a broad survey of hospital managers in German-speaking countries. Although the survey is focused on Germany as the largest healthcare market in Europe the results are considered to be exemplary and generic to other European countries as they reflect crisis scenarios described in the literature [9]. To our knowledge there is no preceding project presenting a systematic evaluation of crisis management in hospitals differentiating the described areas. As a next step the survey is going to be active until the first quarter of 2014 and authors expect to submit results from it for publication in the second half of 2014. Based on the findings of the expert workshops and the expert survey authors plan to develop a decision-support-tool that will extend the capabilities of standard risk assessment methods with the presented holistic and domain-specific approach and thus provide hospital managers with a tailored instrument for crisis mitigation and aversion.

# REFERENCES

1.  Deutsche Krankenhausgesellschaft, Eckdaten der Krankenhausstatistik, DKG, 2013.

2.  B. Augurzky, D. Engel, C. M. Schmidt, and C. Schwierz, "Ownership and financial sustainability of German acute care hospitals," Health Economics, vol. 21, no. 7, pp. 811–824, 2012.

3.  B. Dean, M. Schachter, C. Vincent, and N. Barber, "Causes of prescribing errors in hospital inpatients: a prospective study," The Lancet, vol. 359, no. 9315, pp. 1373–1378, 2002.

4.  G. Armitage, "Human error theory: relevance to nurse management," Journal of Nursing Management, vol. 17, no. 2, pp. 193–202, 2009.

5.  H. C. Ko, T. J. Turner, and M. A. Finnigan, "Systematic review of safety checklists for use by medical care teams in acute hospital settings—limited evidence of effectiveness," BMC Health Services Research, vol. 11, article 211, 2011.

6.  V. Valdmanis, P. Bernet, and J. Moises, "Hospital capacity, capability, and emergency preparedness,"European Journal of Operational Research, vol. 207, no. 3, pp. 1628–1634, 2010.

7.  P. Yi, S. K. George, J. A. Paul, and L. Lin, "Hospital capacity planning for disaster emergency management," Socio-Economic Planning Sciences, vol. 44, no. 3, pp. 151–160, 2010.

8.  T. Hoff, L. Jameson, E. Hannan, and E. Flink, "A review of the literature examining linkages between organizational factors, medical errors, and patient safety," Medical Care Research and Review, vol. 61, no. 1, pp. 3–37, 2004.

9.  M. Dückers, M. Faber, J. Cruijsberg, R. Grol, L. Schoonhoven, and M. Wensing, "Safety and risk management interventions in hospitals: a systematic review of the literature," Medical Care Research and Review, vol. 66, no. 6, supplement, pp. 90S–119S, 2009.

10.  J. Lauterberg, K. Blum, M. Briner, and C. Lessing, Befragung zum Einführungsstand von klinischem Risiko-Management (kRM) in deutschen Krankenhäusern, Institut für Patientensicherheit der Universität Bonn (IfPS), 2012.

11.  V. Stantchev, T. D. Hoang, T. Schulz, and I. Ratchinski, "Optimizing clinical processes with position-sensing," IT Professional, vol. 10, no. 2, Article ID 4476251, pp. 31–37, 2008.

12.  A. K. Jha, C. M. Desroches, E. G. Campbell et al., "Use of electronic health records in U.S. Hospitals,"The New England Journal of Medicine, vol. 360, no. 16, pp. 1628–1638, 2009.

13.  C. M. DesRoches, C. Worzala, M. S. Joshi, P. D. Kralovec, and A. K. Jha, "Small, nonteaching, and rural hospitals continue to be slow in adopting electronic health record systems," Health Affairs, vol. 31, no. 5, pp. 1092–1099, 2012.

14.  A. L. Kellermann and S. S. Jones, "What it will take to achieve the as-yet-unfulfilled promises of health information technology," Health Affairs, vol. 32, no. 1, pp. 63–68, 2013.

15.  C. P. Landrigan, G. J. Parry, C. B. Bones, A. D. Hackbarth, D. A. Goldmann, and P. J. Sharek, "Temporal trends in rates of patient harm resulting from medical care," The New England Journal of Medicine, vol. 363, no. 22, pp. 2124–2134, 2010.

16.  R. Reece, "Why doctors don't like electronic health records," MIT Technology Review, 2011.

17.  V. Stantchev, Enhancing Health Care Services with Mixed Reality Systems, Springer, Berlin, Germany, 2009.

18.  V. Stantchev, "Intelligent systems for optimized operating rooms," in New Directions in Intelligent Interactive Multimedia Systems and Services, E. Damiani, J. Jeong, R. Howlett, and L. Jain, Eds., vol. 226, pp. 443–453, Springer, Berlin, Germany, 2009.

19.  M. Armbrust, A. Fox, R. Griffith et al., "A view of cloud computing," Communications of the ACM, vol. 53, no. 4, pp. 50–58, 2010.

20. M. Jafari, R. Safavi-Naini, and N. P. Sheppard, "A rights management approach to protection of privacy in a cloud of electronic health records," in Proceedings of the 11th Annual ACM Workshop on Digital Rights Management, pp. 23–30, New York, NY, USA, October 2011.

21. BIS, Crisis Management, Department for Business Innovation and Skills, 2010.

22. S. Rivkin and F. Seitel, "Taming a hospital crisis: 7 rules of the road," Becker's Hospital Review, 2012.

23. J. S. Robinson, "A continuing saga of piped medical gas supply," Anaesthesia, vol. 34, no. 1, pp. 66–70, 1979.

24. V. Stantchev and G. Tamm, "Reducing information asymmetry in cloud marketplaces," International Journal of Human Capital and Information Technology Professionals, vol. 3, no. 4, pp. 1–10, 2012.

25. V. Stantchev and G. Tamm, "Addressing non-functional properties of services in IT service management," in Non-Functional Properties in Service Oriented Architecture: Requirements, Models and Methods, pp. 324–334, IGI Global, 2011.

26. E. P. Wright, C. H. Collins, and M. D. Yates, "Mycobacterium xenopi and Mycobacterium kansasii in a hospital water supply," Journal of Hospital Infection, vol. 6, no. 2, pp. 175–178, 1985.

27. U. Beckmann, D. M. Gillies, S. M. Berenholtz, A. W. Wu, and P. Pronovost, "Incidents relating to the intra-hospital transfer of critically ill patients: an analysis of the reports submitted to the Australian incident monitoring study in intensive care," Intensive Care Medicine, vol. 30, no. 8, pp. 1579–1585, 2004.

28. E. J. Anaissie, S. R. Penzak, and M. C. Dignani, "The hospital water supply as a source of nosocomial infections: a plea for action," Archives of Internal Medicine, vol. 162, no. 13, pp. 1483–1492, 2002.

29. M. Best, V. L. Yu, J. Stout, A. Goetz, R. R. Muder, and F. Taylor, "Legionellaceae in the hospital water-supply. Epidemiological link with disease and evaluation of a method for control of nosocomial legionnaires' disease and Pittsburgh pneumonia," The Lancet, vol. 2, no. 8345, pp. 307–310, 1983.

30. G. Hornung and C. Schnabel, "Data protection in Germany I: the population census decision and the right to informational self-determination," Computer Law and Security Review, vol. 25, no. 1, pp. 84–88, 2009.

31. BBK, Schutz Kritischer Infrastruktur: Risikomanagement im Krankenhaus—Leitfaden zur Identifikation und Reduzierung von Ausfallrisiken in Kritischen Infrastrukturen des Gesundheitswesens, Bundesamt für Bevölkerungsschutz und Katastrophenhilfe (BBK), 2012.

32. K. Nagaraju and R. Willmann, "Developing standard procedures for murine and canine efficacy studies of DMD therapeutics: report of two expert workshops on "pre-clinical testing for Duchenne dystrophy": Washington DC, October 27th-28th 2007 and Zürich, June 30th-July 1st 2008,"Neuromuscular Disorders, vol. 19, no. 7, pp. 502–506, 2009.

# Chapter 2

# A Business Process Model of Inspection in Remanufacturing

Mark Errington and Stephen J Childe

College of Engineering, Mathematics and Physical Sciences, University of Exeter, EX4 4QF, Exeter, UK

## ABSTRACT

A crucial stage of the remanufacturing process is the inspection procedure. Surveys carried out in the automotive remanufacturing sector show that the industry is concerned about the need this causes for a large amount of specialist skills. Despite this, there has been little research into what is actually involved in the inspection process and its outcomes.

This paper presents case-based research that was carried out on the inspection procedures of both electronic and mechanical product remanufacturers. It presents generic inspection process diagrams, produced using case studies in UK companies engaged in remanufacturing activities. The models provide a greater understanding of the remanufacturing inspection procedures currently used. The models were tested with additional case studies.

The paper discusses the questions raised by the improved understanding of inspection processes in remanufacturing for operations managers and outlines some questions for future research.

# BACKGROUND

Early results of this research were described in [1]. Due to global warming, the Climate Change Act [2] and the imminent peaking of world oil production (as described in [3] and [4]), closer attention is being paid to the impact of manufacturing and its energy consumption on the environment. There is now an increasing amount of legislation being developed by the European Union aimed at reducing energy use as well as our impact on the environment. As more and more countries join the European Union, more manufacturers will be required to comply with this legislation. This includes the so-called producer responsibility legislation including the WEEE directive [5], End of Life Vehicles Act [6], Packaging and Packaging Waste Directive [7] and the Batteries and Accumulators Directive [8]. This legislation requires the producers of certain products to be responsible for their disposal at end of life.

Producers are not only responsible for retrieving the used products from their end users but are also responsible for recycling them whilst meeting certain minimum recycling rates. With limited room remaining in landfill sites, material recycling is often preferable to land filling; however, most of the value added during manufacture and the energy embodied in the product is lost.

In order to take back post-consumer products, it is necessary to create a reverse supply chain. In cases when this is integrated into forward manufacture, in such a way that old products are used to produce new, at any level, this leads to a 'closed-loop' supply chain [9].

Many consumer products are still working at their end of life, and there are an increasing number of companies that collect these products and return them to market both in the UK and abroad. They provide a cheap alternative to buying new products and appear to have been very successful both in not-for-profit and for-profit organisations.

Other products reach their end of life because of an accumulation of rectifiable faults, the wearing of a small number of components within a product or sometimes because of their age or amount of use. It has been shown that there has been a long tradition of remanufacturing in the auto parts industry [10]. Manufacturers in other industries such as Hewlett Packard [11], Océ [12] and Xerox [13] have started to take advantage of this way of doing business and have started to run large remanufacturing operations. It is often stated that remanufacturing is 80% more energy efficient and 60% more cost efficient than traditional manufacturing. For this reason, a large amount of attention is being paid to these types of processes by old and emerging economies alike.

Many tools now exist which can be used to evaluate the environmental and financial performance of a given product against set criteria [14-17]. The aim of these tools is to identify areas of the design that could be changed in order to enable end of life strategies to be carried out more efficiently.

There has also been a large amount of research carried out on operations management within the remanufacturing process itself. Guide et al present an analysis of the performance of different static priority rules [18], while in an earlier paper, describe the application of Goldratt's drum-buffer-rope system in a remanufacturing environment [19,20]. Other authors look at the use of production planning approaches in remanufacturing environments [21-31].

Reverse logistics channels themselves have also attracted a large amount of interest. Jayaraman et al. present a deterministic integer programming model which can be used to determine the optimum location for remanufacturing facilities and collection centres [9]. A similar optimisation study has been carried out by Peng ZY and Zhong [32]. Blackburn et al. discuss the need for different reverse supply chains for different models [33]. They find that depreciation rates among high-value items mean that responsive supply chains are more appropriate in some cases.

There are also some studies that look at strategies for increasing the number of cores (end-of-life items) available for remanufacture. Ray et al. present a method for determining the correct buy-back pricing strategy for used products from consumers [34]. Xiaochen et al. present a method for calculating the optimum buy-back price under certain conditions [35].

# INSPECTION AND DISPOSITION DECISIONS

'Inspection and disposition' is one of the five main parts of a remanufacturing business [36]. Despite this, research has tended to focus largely around product design for remanufacture rather than upon the operational questions related to how a specific product is actually inspected and remanufactured. Steinhilper [37] identifies that:

"(A) step of great importance in remanufacturing is to assess the condition of the disassembled and cleaned parts as to their reusability or reconditionability."

He goes on to state that this is done in two parts: to define the objective criteria and accepted condition characteristics and to determine how this will be assessed.

Guide and Van Wassenhove [36] describe the inspection and disposition process as follows:

"The testing, sorting and grading of returned products are labour-intensive and time-consuming tasks, but the process can be streamlined if a company subjects the returns to quality standards and uses sensors, bar codes and other technologies to automate tracking and testing."

There are many signs that the technology that is referred to in this text is beginning to be more widely used. Bosch have carried out tests with micro sensors to record data during the life of its power tool motors [38]. Their plan was to connect the tools to a data reader once they have been returned to them in order to decide whether the motors can be reused or not. At the time, similar technology was being developed as part of the Care Vision 2000 initiative to produce what they call a 'Green Port' [39].

Technology to make the inspection and disposition decision more streamlined is clearly useful in some cases. However, it is not applicable or relevant for many remanufacturing operations. These techniques are only useful to those companies who originally manufactured the product or have control over its design. Companies wishing to use such technology would have to invest with a view for a very long-term payback.

Inspection and disposition is carried out in all remanufacturing operations; however, little has been done to document current industry practices, to identify areas where savings could be made and to assess exactly which specific inspection tasks these technologies should be substituted for.

In contrast to new-product manufacture, where sampling methods are often used, remanufacturing always requires 100% inspection [40]. This is typically carried out on cores and disassembled parts as well as the finished products themselves. This is done to increase the second user's confidence in remanufactured products, and it is thought to explain why remanufactured products appear to have a better reliability than new products [40].

A comprehensive survey of manufacturers in the automotive industry was carried out in order to identify the problems the industry faces from the perspective of the remanufacturers themselves [10]. Twenty-nine percent of respondents stated that the factor which made inspection most difficult was the knowledge of the employee carrying out the work. A further 21% of respondents stated that identifying defects in cores was their main concern.

This suggests that there is a need for formal methods that can be used to simplify the processes involved in the inspection procedure. This would remove the dependence on such a high level of tacit knowledge held by the operators. As noted above, some attention has been focussed on the aspects of design that can facilitate remanufacture. In the survey, the authors note that design for remanufacturing is an issue with automotive remanufacturing [10]. 'Profit potential' and 'investment needed to repair' rank high as responses to how the decision to remanufacture a given product is made.

# RESEARCH OBJECTIVES

Researchers have established that inspection is one of the key stages in remanufacturing. The main aim of the inspection process is to determine the condition of the returned item (called the 'return' or the 'core') and select the most economically attractive re-use option [11]. Little has been done to look at the tasks involved in this process and how they can be carried out more efficiently. This paper describes work done to address this issue through case study research. It aims to establish how these processes are currently carried out as a basis for understanding and to provide a basis for analysis of potential improvements. Using the four purposes of research described by Voss et al. [41], the work is theory building since it attempts to develop new models or concepts that can help researchers and managers understand and deal with inspection. Using a broad range of cases involving different products, industries and business models but all centred on remanufacturing, the research was able to take a broad, general understanding of the issue which can be further tested and refined by future work.

The main objective of these case studies was to gain a detailed insight into the role and process of inspection operations carried out by remanufacturing firms. Despite the fact that research in this area is very immature, it is suspected that there is a huge quantity of technical expertise within the companies that in some cases have been carrying out these activities profitably for many years.

The work aimed to establish the objectives of inspection and testing at all stages of the remanufacturing process and to investigate the possibilities of gaining the same process outputs through different means.

# METHODOLOGY

Case research is a very useful technique for carrying out work in immature research areas [42]. It has also been established that it can be used effectively where there is a large amount of expertise in the field that needs to be captured and formalised. Eisenhardt [43] shows how theory can be generated using case research and how it is a particularly useful method in new topic areas. Scientific rigour was ensured through following the framework for case research described by Voss et al. [41].

Discussion points were drawn from the literature and further developed using the findings from these cases. These included boundary questions to understand the characteristics of the case companies that were being studied. These included asking if the company was the original equipment manufacturers (OEM) of the products which it remanufactures and how diverse the mix of products remanufactured by each company was. The interviewees were asked to describe a remanufacturing process within their organisation and the role of inspection within it.

Cases were selected so that a broad range of industries were represented. These include OEMs and non-OEMs as well as for-profit and not-for-profit companies. Case studies were carried out with a number of companies engaged in remanufacturing in the UK. These operate different business models for a wide range of customers. Some carry out remanufacturing on a not-for-profit basis, and others remanufacture directly for OEMs on a large scale. The table gives some of the characteristics of the different companies that were included in the research.

It can be seen from Table 1 that the companies studied were from a wide range of industries with different products and different business objectives. The only not-for-profit organisation, CompCo, operates to give people in developing countries access to IT equipment. This adds an interesting characteristic since many of their staff work as volunteers, although this does not mean that labour is a cheap resource that may be used inefficiently. The remainder of the companies operate in a more conventional for-profit way. The prominence of automotive and defence sector remanufacturers in the study reflects the maturity of remanufacturing in these areas.

**Table 1:** Overview of case study companies

| Company identifier | Customer type | Sector | Product type studied | OEM? | For profit? |
|---|---|---|---|---|---|
| DefCo | UK Ministry of Defence | Defence equipment | Mechanical/ electronic | OEM approved | For profit |
| CopyCo | Business | Copying/ Printing equipment | Mechanical | OEM | For profit |
| PCCo | Business | IT equipment | Electronic | Non-OEM | For profit |
| GearCo | OEM | Automotive | Mechanical | OEM Approved | For profit |
| ClutchCo | Business | Automotive | Mechanical | Non-OEM | For profit |
| CoreCo | Remanufacturers | Automotive core broker | Mechanical/ Electronic | Non-OEM | For profit |
| TurbineCo | Business | Power generation | Mechanical | OEM | For profit |
| CompCo | NGOs | IT equipment | Electronic | Non-OEM | Not for profit |
| MilCo | UK Ministry of Defence | Defence equipment | Mechanical | OEM | For profit |

Errington and Childe

Errington and Childe *Journal of Remanufacturing* 2013 3:7
doi:10.1186/2210-4690-3-7

'Product type studied' describes the remanufactured component, module or product that was studied for this research. This is not necessarily the main product the company produces. Some of the cases remanufacture a large volume of products and operate in a highly competitive market while others remanufacture smaller numbers of items and are known for their expertise and competency.

Many of the cases studied were found not to describe their process as remanufacturing. For the purposes of this research, the process was considered to be remanufacturing if it met the definition developed by Ijomah and Childe [44]. This is a definition that is widely accepted in the UK remanufacturing industry. It defines a remanufacturing process as follows:

"Remanufacturing is the only process where used products are brought at least to OEM performance specification from the customer's perspective and, at the same time, are given warranties that are equal to those of equivalent new products."

Companies were approached through the UK government-funded Centre for Remanufacturing and Reuse and through university contacts. The authors note that other definitions for remanufacturing exist, and a full discussion can be found in the PhD thesis Business Processes.

The main difference between this definition and others is its inclusion of the word 'warranty'. The authors of this definition felt this was vital as anyone claiming that their remanufactured product was as good as or better than the new but only willing to offer a shorter warranty may not be particularly confident in their performance claim.

Company visits were carried out in mid-2006, these included a meeting with the operations managers as well as the production staff followed by a tour of the facilities being used for remanufacturing. The purpose of these visits was to gain an overall understanding of the inspection processes carried out in each of the case companies.

Complete notes were compiled directly following each visit. These consisted of background information about the company and a description of the remanufacturing and inspection processes. These included flow charts of the processes that were studied. These notes were emailed directly to those who were interviewed, for checking. The purpose of this was to ensure their internal validity and correctness. Results from case studies were triangulated through visits to the factory floor and further interviews with the production staff. After each visit, the case notes were sent to the interviewees promptly to verify their accuracy.

An analysis of the data collected through the case study meetings is presented in the following section.

# RESULTS AND ANALYSIS

In this section, we present four of the cases used in the research. Two are from the automotive industry, and two are from the computer industry. The cases show that there are similarities and differences within each industry and between the industries. It will be shown later how a general model can be structured to cover all the options.

## GearCo

GearCo is a remanufacturer of automotive drive trains. They remanufacture engines, transmissions and gearboxes as an OEM approved remanufacturer.

Used items are stored when received, until the arrival of an order for that type of item. When an order arrives, an initial visual inspection of the core is made for obvious damage which would make the core uneconomical to repair, for example a cracked casing. The reason for failure of a returned core is not always obvious, and is not taken into consideration. The core is disassembled and cleaned using a three-stage cleaning process. A glass bead blast is used as well as high- and low-pressure chemical cleaning.

Some parts are always replaced, such as bearings or other wear items. These are removed and scrapped. Other parts are inspected to assess their potential for re-use. Experienced inspectors assess parts against criteria set in conjunction with the OEMs. This decision is made by taking measurements of the item and assessing the amount of wear it has experienced. This can be used to estimate the amount of life remaining in the part. For these products, there are no parts at all that are never replaced.

Some of the smaller parts are removed from the core and sent through a separate cleaning and inspection procedure. The main parts are placed in a tray together. An inspector uses a computer-based system to establish the part numbers of the replacement parts that will be required for the unit and uses the system to generate a stores order for them. It is at this point the failed parts are removed and scrapped.

All the replacement parts that are used are supplied by the OEMs or OEM-approved suppliers. Parts are never repaired. It would be possible for non-OEM parts to be used, but then the remanufactured gearbox would not be allowed to carry a full OEM warranty.

Once assembled, the unit is put through a testing procedure developed in conjunction with the OEM. The gearboxes are run both with and without load and analysed for noise and operation. Acceptable limits are set in conjunction with the OEM. One hundred percent of the units produced are inspected in this way.

Those units that fail this procedure are sent for rework by the operative who was responsible for assembling the unit.

## ClutchCo

ClutchCo supplies 1,300 types of clutches used in cars in the UK and Europe. Half of these models, by part number, come from remanufacturing, and half are new imports, coming from China. The remanufactured clutches account for half the product lines but only 20% of sales volume. This strategy of using remanufacturing to supply lower volume items allows the company to hold less stock of slower moving parts at the same time as allowing it to have full market coverage. Thus, the imported clutches account for 80% of the total clutch sales, and these are cheaper. ClutchCo accepts that its remanufacturing area may now be running at a loss; however, it estimates this cost to be less than the cost of keeping the large stocks of slow moving parts that would be necessary if it was to import them from China.

As an independent remanufacturer, ClutchCo has had little cooperation from the OEMs of the products it remanufactures. This is despite the apparent legal obligation of the OEMs to cooperate. (The End of Life Vehicles Directive requires all information required for the correct and environmentally sound treatment of end-of life vehicles to be made available to authorised treatment facilities by vehicle manufacturers and component producers [6]. In order to establish their processes for remanufacturing, ClutchCo purchases OEM clutches and reverse-engineers them. They have found this to be a more effective method than obtaining information from the OEMs directly. None of the clutches that are remanufactured by ClutchCo are supplied to OEM manufacturers.

After the cores arrive at the plant they are sorted. They are identified by part type, against approved samples, and are inspected visually for any obvious damage that would make them unusable. The main fault inspectors are looking for is excessive wear to clutch diaphragms. Excessive wear would mean that the reassembled clutch would be unlikely to last a full second life. After inspection, the cores are formed into production batches and are stored in the warehouse.

When required, cores are collected from the warehouse and washed using a caustic solution. Sometimes the clutch diaphragm is tested before disassembly. The cover of the clutch is removed to facilitate access and all parts are cleaned in order to make them appear as new. Inspection of the clutch's component parts is carried out. Parts are worked upon or scrapped as necessary.

The clutch is then reassembled and tested for clamp load and clearance. Clutches that fail this test are reworked. Finally each clutch is ink-jet sprayed with its part number and batch code.

# PCCo

The mission of PCCo is to extend the life of office equipment. They offer a wide range of services ranging from collection and auditing of unwanted equipment to the supply and installation of refurbished equipment. The majority of the work undertaken is to dispose of used IT equipment. Where possible, they do this through remanufacture and resale of the equipment.

Items that are deemed to be unmarketable are separated from rest of the load of incoming equipment and are sent directly to a recycler. At the time of study, this included 15" CRT monitors.

The remaining items are sent to the remanufacturing facility. On arrival, a batch number is allocated to the equipment, and a unique tracking number is given to each piece of equipment. Information regarding the type of equipment, manufacturer and model number is stored against the tracking number at this stage.

The batch is then sent to the workshop where it is processed. Equipment is connected to a network which automatically audits and tests it. The network stores data about the equipment, carries out data erasure in accordance with the UK government standards and creates a record of the wipe. Software is used to carry out a fault diagnosis of the equipment.

The majority of equipment is found to be working. The exact amount varies depending on its source, but the average is similar to the 80% figure quoted for Germany by Steinhilper [37]. A visual inspection is made, and the equipment is cleaned as necessary.

Information collected during the diagnostic 'bench' test is used in order to determine the potential for remanufacturing each item of the equipment. This is restricted by value, functional state and age. A triage principle is used to sort the equipment into three streams. These are 're-use', where the computer is reused without further treatment, 'recycle', where the computer is sent for materials recovery and 'process further', where the faults with the computer are not known but it may be possible to reuse after further operations. Diagnostic information, collected during the bench test, is then used to estimate the parts and labour cost of processing each item of equipment according to its individual requirements.

If the decision is made to process an item further, then the item is disassembled and an investigation is made into the cause of the equipment failure. Once the cause is identified, this module or part is replaced with a working one, and the equipment is reassembled An electrical safety test is carried out then each computer is packaged for sale and shipping.

# CompCo

CompCo remanufactures between 1,400 and 1,700 computers per month. The vast majority of these are desktop computers, but they also process a smaller number of laptop computers. CompCo works on a not-for-profit basis and provides computers to educational establishments and NGOs in various locations throughout the developing world. The standard processing lead time is 2 weeks, and stock does not spend much time in the warehouse.

Computers are sorted at source as much as possible. Only high specification, working machines are accepted. Computers that are found not to work or to be the wrong specification are recycled with the cost, plus a premium charge for disposal, borne by the donor. This is meant to incentivise donors to supply the required machines, rather than use CompCo for general disposal.

Computers arrive at the warehouse in an unknown state. Information on the condition of the computers is sometimes given by the donor, but it usually turns out to be inaccurate. For this reason, the decision making about the suitability of a machine for remanufacture is made entirely on the basis of a visual inspection, the model of the computer and an estimation of its age. This is used to estimate the length of usable lifetime a computer will have once it is fully working. If it is decided that a computer is not suitable for remanufacturing it is sent for part recovery and/or materials recycling.

During the next stage of the process, the data wiping, the machine specifications are found using a diagnostic software. Based on this, it is decided if the computer is suitable for reuse as it is or if it requires upgrading or part replacement. If the equipment cannot be used, then it is sometimes cannibalised for spare parts.

No standard part replacements are made but computers are cleaned and tested for safety before they are shipped.

# Generic Inspection Procedures in Remanufacturing Processes

It was found that the processes undertaken by the remaining cases were very similar to those described above. Electronic product inspection and remanufacture was carried out using the same steps at CompCo and PCCo whereas mechanical product inspection was carried out using the same steps at GearCo and ClutchCo. All of the processes were found to have certain stages in common.

The flow charts that were generated during each of the case studies were compared, and common stages of the process were extracted. Focussing particularly upon inspection processes, a name was selected for each stage of inspection that was common across the cases. This led to the creation of a model that could be described as a 'common ground' or 'consensus' model which applies across the cases. This could form the basis of a generic inspection procedure.

The model was shown graphically using the IDEF0 modelling standard. The IDEF0 standard was used in preference to other modelling techniques such as flow charts because it allows decomposition, is easy to understand due to the simple graphics that are used, is precise and can be used with data abstraction. This is a particularly useful feature in the creation of a generic process diagram such as the one described in this paper. Mechanism arrows were not included in the diagrams as these are often specific to each company. A company might choose to use a machine to carry out a given inspection whereas another might carry out the process using skilled labour. It is the activity or job that must be carried out that was of interest for this research rather than the exact way it was carried out in a specific company. For a full discussion of the benefits of the IDEF0 standard in process modelling, see [45] or [46].

Figure 1 shows the generic remanufacturing process that was observed in the cases.

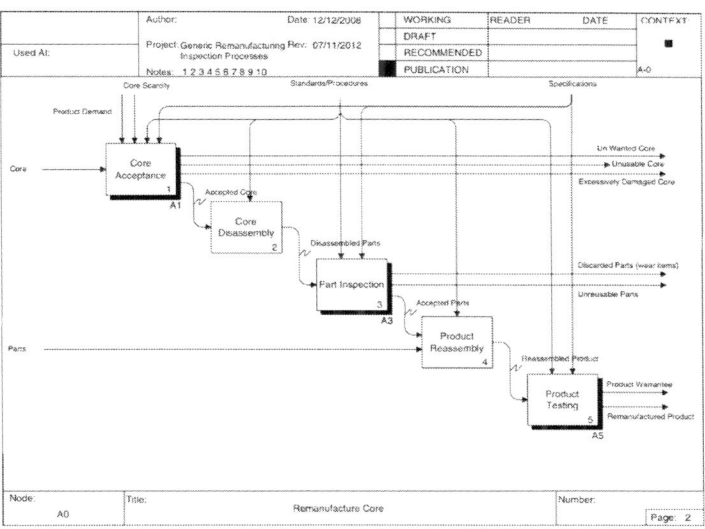

**Figure 1:** IDEF0 diagram of generic remanufacturing process.

It can be seen that three key areas of inspection were found. Each was carried out at a different stage in the remanufacturing processes. These were identified as follows:

core acceptance;

- part inspection; and
- final product testing.

The different stages of inspection have different objectives. Due to these different objectives, the physical inspection processes themselves are different. The objective of *core* inspection and testing is to remove cores that will be uneconomical or impossible to remanufacture, accepting only those thought to be viable. This improves the reliability of the population of items that are produced and ensures that cores that are uneconomical to remanufacture do not enter the process. This process is often carried out by a triage-type process. This was referred to as 'scratch and sniff' by an expert in the automotive remanufacturing industry.

The second stage of inspection is carried out once the core has been disassembled. *Part* inspection and testing aims to remove non-reusable components from the product in order to increase its reliability once it is reassembled. These parts may be non-reusable because they have already failed or because they are likely to fail within the next working lifetime of the product. This is typically done using a visual inspection for wear, but in other processes, measurements are made using lasers, and some parts are tested for performance.

The *final product* inspection stage is carried out to ensure that the products are in full working order before they are shipped. Products that fail this stage are reworked before being retested and sold. This process is often similar to typical final product testing in traditional forward manufacturing. Products are operated and their performance assessed to ensure they are within acceptable limits.

All the inspection and testing procedures are carried out in order to increase confidence in the ability of a product to perform throughout its second life. Without these procedures, it would be difficult to ensure the quality of the products produced and to offer full 'as new' guarantees as required for remanufactured products. This could potentially lead to a lack of consumer confidence in remanufactured products and excessive guarantee costs to the remanufacturer.

The effect of the three stages of inspection was to remove from the process those items that were of less value either because of requiring more work or being more likely to fail in service. In some cases, these items were not discarded but were disassembled in order to reuse good components.

# Core Acceptance

Cores are usually inspected as soon as they arrive at the remanufacturing facility or at the first stage in the process once items are taken from storage for use. This is the process which Loomba proposes should be used more to reduce remanufacturing costs [47]. Figure 2 shows the details of how this inspection is carried out.

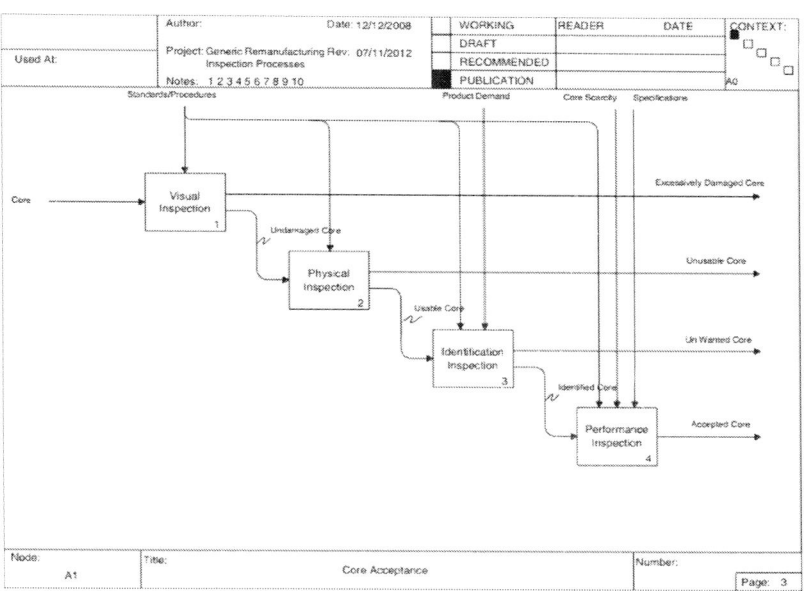

**Figure 2:** IDEF0 diagram of generic core inspection procedure.

The figure shows the four main tasks that were identified in the core acceptance procedure. The first is a visual inspection. This is carried out simply through looking at a core and deciding whether it has obvious major damage which would make it unusable. An example of products that would be rejected at this stage would be products that have been crushed during transit to the remanufacturing facility. All the cases used visual inspection with the aim of removing excessively damaged items from the process. However, the case of CompCo suggests visual inspection could be entrusted to suppliers given certain conditions.

Cores that have no obvious signs of major damage are subjected to the second test in the procedure. This has been described in Figure 2 as the physical inspection. Again this is done manually, but the actual process varies depending on the core being inspected. For automotive components, two main methods are commonly used. The first is to attempt to rotate any part of the product that should normally move and the second to smell electrical components to test for burn out. Cores that fail this test are sent for recycling and/or disposal.

The third stage of the process is to identify the part type and part number of the core. This is used to estimate the demand for and value of the product after it has been remanufactured. Cores for which there is no demand are disposed of at this stage. It may seem counter-intuitive that the identity of the part is only investigated at this stage of the process. One might assume that it would be sensible to identify a component before any analysis is made of its potential for remanufacture. The reason that was given for the identification belonging to this stage of the process was the relative difficulty of the first three stages of inspection. To carry out a visual and physical inspection, known by CoreCo as 'Scratch and sniff' is a very quick process whereas identifying a core through the use of part numbers and expertise can be relatively time consuming. This was stated by CoreCo as being particularly the case for steering columns which do not typically carry part numbers and must be identified by highly trained operatives. This identification operation allows inventory to be managed for future use.

The fourth and final task within the core acceptance procedure aims to assess the performance of the core. This is usually done using test rigs specifically designed for the procedure. The standards the product must meet are set by OEMs, industry bodies, international standards or the remanufacturing firm itself. The aim of this process is twofold: firstly, to establish whether the core is economical to process and secondly to establish whether the finished product is likely to conform to specifications once it has been remanufactured. This is another opportunity to remove unusable items from the process, occurring once simpler checks have been carried out.

## Part Inspection

Once cores have passed the core inspection and testing procedure, they are sent for disassembly and further inspection. After disassembly and cleaning, they are passed to the second set of inspections and tests. The diagram for this process is shown in Figure 3.

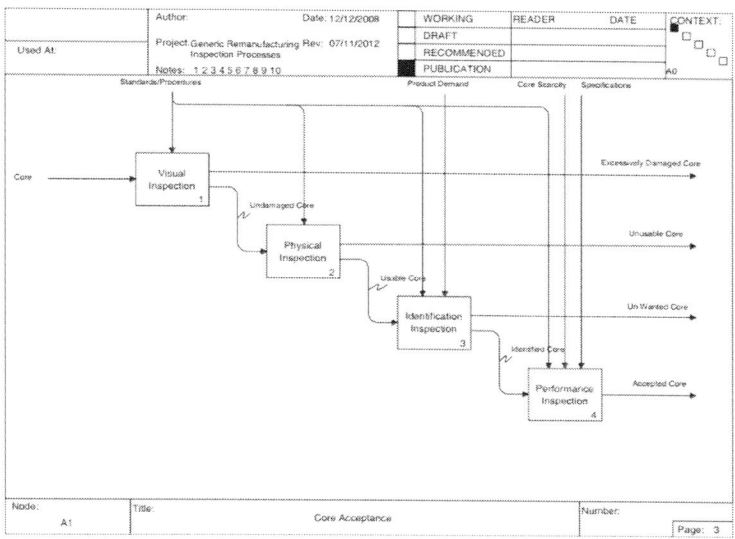

**Figure 3:** IDEF0 diagram of generic part inspection procedure.

It can be seen from the diagram that some of the tasks carried out on the parts are similar to those carried out on the core. Firstly each part is identified. If, according to company procedures, it is a part that is always replaced, then it is discarded immediately. These are usually wear components, such as bearings and bushes, and are often the reason for previous user discarding the product. When parts may be replaced, identification is clearly necessary.

Parts that could be reused are sent through a functional inspection process. Typical activities in this process include measurement and leakage testing. If they are found not to conform to required specifications, whether for functional or safety reasons, then they are either discarded or reconditioned. It was found that this decision was mainly made on the basis of cost and lead time.

During the case study interviews, it was stated that it is often quicker to recondition a part in house than it is to order a new one to be made. A similar comment is made by Depuy et al. [48]. This is especially the case when small numbers of parts which are not mass produced are concerned. In practice, this seems to make the reconditioning of parts a part of the inspection process, wherein the components are inspected until they are accepted or failed. This contrasts to manufacturing, where parts are normally processed until they are finished, with inspection as a final operation, and makes the remanufacturing process more similar to craft work or 'fitting'. After reconditioning, parts are inspected once more for performance in order to ensure they meet requirements.

It was observed at CompCo that no physical disassembly was ever performed, so physical parts were never assessed or replaced. In this case, the disassembly consisted only of removing data and reassembly only or installing of new software. Nevertheless, this case was regarded as remanufacturing (rather than re-use) since the confidence gained from the inspections and tests allowed the company to offer OEM level specification and warranty.

Once the quality of parts has been accepted, additional parts, either new or reconditioned from other sources, are used to replace the ones discarded in the previous inspection stage. The product is then reassembled. Once it is reassembled, it is passed to the third and final stage of the inspection procedure, the product testing stage.

## *Product Testing*

It can be seen that functional inspection is the key part of this inspection and testing process. This is the final stage at which the reliability of the final product can be estimated and/or ensured. The diagram for this process is shown in Figure 4. The process carried out and the standards applied during the functional inspection are different for each product. In the case of OEM-approved remanufacturers, these are developed with involvement of the OEM for the product concerned. This ensures satisfactory operation.

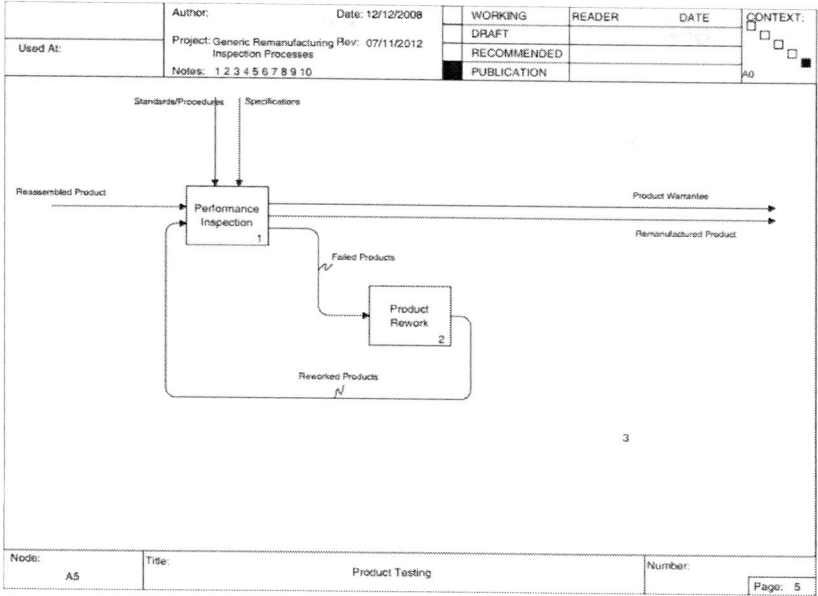

**Figure 4:** IDEF0 diagram of generic final product inspection procedure.

Products that fail the functional inspection are reworked and retested before they are sold. It would seem logical that products would be sent back to the disassembly process to be disassembled (where this involves a dedicated area or workplace). However, in practice, it was found to be more common for the product to be sent back to the person who reassembled it. GearCo stated that this was done because the re-assembly operative would be likely to know what the cause of the fault would be. Thus, the corrective actions are again more closely linked to the inspections than in manufacturing.

# DISCUSSION

There appear to be two motives for inspection: (1) to remove unreliable individuals from the population and (2) to show that enough work has been done to achieve the required level of performance.

By removing unreliable individuals, the confidence in the remanufactured products is raised. However, this does not mean that all products require all the possible inspection stages. The processing that units go through will be identical for all the units that are not obviously damaged or in a doubtful condition. It is simply not necessary to know the details of the state of a unit. By performing the simplest and cheapest inspection operations first, the parts that are removed by these operations do not undergo the more expensive or time-consuming activities. This kind of inspection could be termed inbound inspection and consists of identification and some kind of simple assessment of condition. Inbound inspection is not concerned with the units as individuals but as feedstock. This applies at the core level and at the part level.

By showing that enough work has been done, the second type of inspection becomes part of a cycle where products go through a loop of testing, re-work, re-assembly, testing until the performance is found to be adequate. At first glance, this looks like trying to 'inspect quality into a product' which Deming was against (Dodge quoted in Deming 1982 [49]). However, Deming was looking at the whole system of manufacturing, in which quality depends on product design and production system design. In this instance, it seems better to think of this cycle of inspection as adjustment or tuning of the product. For example, balancing is a normal requirement in the manufacture of rotating machines, and the activity of balancing continues until acceptable balance is achieved. The balancing cannot improve the quality of the components, but it provides correct performance of the product. This could be termed outbound inspection and is concerned with the units as individuals, providing specific attention to the condition of each one. This applies at the part level and at the final product.

It was found that, in most cases, inspection was carried out three times during the remanufacturing process. Although Brent and Steinhilper state that it is always necessary to carry out 100% inspection during remanufacturing processes [40], these cases show that this does not mean all inspection operations are done on all items. The key thing to consider is if the inspection will actually lead to any decisions being made. It is possible that without disassembling a core, sufficient information cannot be gathered to make a confident decision about the functionality of the part. In other cases, due to a shortage of cores, all items have to be disassembled anyway so there is no need to carry out the first inspection.

Some authors propose that information from green port systems or from the user directly can be used in place of some of the stages of inspection. These are particularly applicable to the core acceptance procedure but rely on accurate and reliable information being available. One way to ensure it is available is to lease rather than sell equipment to the customer. It has been widely reported that this has been a particularly successful method for Xerox Corporation.

In the case studies, no major differences between the way OEM and non-OEM remanufactures operated their inspection procedures were observed. OEM remanufactures are thought of as having fewer problems sourcing cores and information with regards to the design of the products. There is no doubt that these make the remanufacturing business easier to run; however, no impact was observed on the inspection procedures of the companies that were studied.

This section of the paper has presented a generic inspection procedure developed from case studies carried out with companies carrying out remanufacturing operations.

The following section will give details of how the models of the procedure that have been presented in this section were validated in order to gain confidence in their generalisability.

# MODEL VALIDATION

Validation of the model was carried out using a group of six companies contacted through a consultancy firm specialising in remanufacturing. The consultancy firm in question is also responsible for running the UK government-funded Centre for Remanufacturing and Reuse. A visit was made to each of these remanufacturers during the late part of 2008, and a presentation was given describing the model, its origin and its potential uses. The practitioners were then asked to fill in a short questionnaire. Responses were recorded on paper. Interviewees were invited to comment on how they thought the models should be altered, where applicable, to show their processes more correctly.

Table 2 shows the companies that were studied for the validation stage of this research.

**Table 2:** Case studies used for validation

| Company identifier | Product | Electronic/ Mechanical |
|---|---|---|
| CompressorCo | Compressors | Mechanical |
| VacCo | High vacuum pumps | Both |
| WasteCo | Televisions (also processes all types of WEEE) | Electronic |
| FlightCo | Aerospace flight control units | Mechanical |
| TonerCo | Toner cartridge | Mechanical |
| AltCo | Rotating automotive electrics | Mechanical |

Errington and Childe

Errington and Childe *Journal of Remanufacturing* 2013 3:7 doi: 10.1186/2210-4690-3-7

It can be seen that the cases were from a variety of industries which remanufacture a large variety of products. WasteCo is relatively new to remanufacturing, and at the time of writing, its processes were undergoing extensive development.

Interviewees were asked to respond to the statement *"The processes that have been described in the presentation closely match what is done in our process."* Each interviewee was asked to place their opinion on a six-point Likert scale. A score of 1 indicated that the interviewee strongly disagreed with the statement, and a score of 6 indicated that the interviewee strongly agreed. There was no neutral point. Table 3 shows the responses of the interviewees for each of the cases.

**Table 3:** Validation results

| Company identifier | Role of interviewee | Remanufacturing volume units/ month | Close match to process (1 to 6 Likert) |
|---|---|---|---|
| VacCo | Global support manager remanufacturing | 4,300 | 6 (Strongly agree) |
| WasteCo | Chairman | 4,000 | 5 (Agree) |
| CompressorCo | Marketing admin | 200 | 5 (Agree) |
| FlightCo | Quality manager | 2 to 3 | 4 (Slightly agree) |
| TonerCo | Production manager | 700 | 5 (Agree) |
| AltCo | Managing director | 10,000 | 5 (Agree) |

Errington and Childe

Errington and Childe *Journal of Remanufacturing* 2013 3:7 doi: 10.1186/2210-4690-3-7

It can be seen that all the respondents agreed that the model closely matched what was done in their processes: 1 case strongly agreed, 5 agreed and 1 case slightly agreed. Overall, there was agreement among validation case interviewees to the view that the models matched what was done in their respective processes.

# CONCLUSIONS

This work has identified the inspection operations used in nine case studies. The activities were grouped into a generic or common-ground model structured into three main stages: core acceptance, part inspection and final product testing. The model was subsequently assessed for its general applicability by representatives of a further six remanufacturing companies. The key contribution of this paper is a deeper understanding of the role inspection plays within remanufacturing processes. It can be used by practitioners to design a plan for remanufacturing their products, and it can be used by researchers as a framework for further research. It shows clearly where inspection technology may benefit the industry and how.

Another key finding was that in contrast to forward or traditional manufacturing, sample inspection is not usually used in remanufacturing processes. The state of the cores entering the system varies widely, each therefore needs to be inspected at the core level, part level or both to ensure that the components that are remanufacture are of suitable quality. One hundred percent inspection is potentially not necessary at the final product stage; however, this is almost always done. A possible reason for this is to increase confidence in remanufactured products.

The cases provide some insight into the use of inspection. Core acceptance procedures aim to remove unusable cores from the process, and the simplest inspections are used first so that most cores are rejected using the cheapest inspection procedure. Identification inspection is not generally the first activity in this stage since it can be time consuming. Functional inspection is generally left until last as it would be uneconomical to carry out this test on cores that could be rejected by cheaper tests. Core acceptance ensures, as far as possible, that only useful items proceed to the next stage, thus reducing the work required at subsequent stages, and removing problem items from the process, thereby improving the reliability of the population of items.

Part inspection is used to identify those parts that need replacing and to assess and re-condition others. At this stage, there is a transition from accepting/rejecting items entering the process to specific assessment of individual items and the work required by each.

The final product inspection stage also sees products as individuals, the work done being tailored to achieving the correct performance from each unit. The activities of re-work and adjustment are only carried out on those units made from good cores, with new or reconditioned parts that have been re-assembled and from which fully acceptable performance can be expected.

The model that has been presented, although based on a limited set of cases, provides a structure that can be used to plan future remanufacturing operations. Some questions for the management of remanufacturing operations could be derived as follows: as operations managers might find these hard to answer, they may also be thought of as research questions.

# Core Acceptance

- What are the criteria for accepting a unit as a useable core (rather than as a source of parts or materials)?
- Which of these criteria can be most readily identified by the simplest inspection techniques? These should be carried out before the more complex, expensive or resource-intensive inspections.
- What are the relative advantages of using an item for remanufacture compared with the value of the parts it contains?

# Part Inspection

- Is dismantling necessary to confirm the acceptable condition of parts? Electronic components (such as in the IT cases) could be assessed by functional tests of the final product.

- What is the trade-off between the cost and energy saving of re-using a part compared to its expected reliability? If an item is part worn, its lifetime (and the lifetime of the product) can be assessed and related to the cost of a new part. Safety- and performance-critical parts may be replaced as a matter of policy without the need for inspection.
- What are the relative benefits and costs of re-conditioning parts in-house, subcontracting the reconditioning or making or purchasing new items?
- What are the inventory ramifications of storing large numbers of used parts?

# Final Product Testing

- What degree of inspection and testing is required considering a product has been made up from known good parts?
- What is the relationship between assembly and adjustment for a particular product family? Should the adjustment form part of the same operation?
- It is to be hoped that the models presented can be developed both by practitioners designing and managing remanufacturing operations and by researchers making further investigations in the area. Some research questions arising from this work are the following:
- Does the framework of core acceptance, part inspection and final product inspection apply across all types of remanufacturing operations?
- What activities may be added or deleted for particular technologies or market situations?
- What is the potential for incorporating remanufacturing activities into conventional manufacturing facilities?

The key contribution of this paper is a tool which can be used for further investigation into the area of inspection in remanufacturing processes within academia and consultancy. It also provides a framework to aid companies in the development of new remanufacturing processes.

# AUTHORS' CONTRIBUTIONS

ME carried out the research and wrote the first and subsequent drafts of the paper. SC supervised the research work and case studies and reviewed, made suggestions and alterations to sections of the paper. Both authors read and approved the final manuscript.

# ACKNOWLEDGEMENTS

The authors wish to thank the many industrial representatives who allowed their operations to be studied and who contributed to the research.

# REFERENCES

1. Errington M, Childe SJ: The Need for Inspection in Remanufacturing Operations. In *The Third World Conference on Production and Operations Management*. Tokyo, Japan; 2008.

2. Rooker L, Benn H: *Climate Change Act 2008*. HMSO, London; 2008.

3. Richard CD, Walter Y: Encircling the peak of world oil production. *Nat. Res. Res.* 1999, 8:219-232.

4. Bentley RW: Global oil & gas depletion: an overview. *Energy Policy* 2002, 30:189-205.

5. EU: *Directive of The European Parliament and of The Council on Waste Electrical and Electronic Equipment (WEEE)*. European Parliament, Brussels; 2003.

6. EU: *Directive of The European Parliament and of The Council on End-of Life Vehicles*. European Parliament, Brussels; 2000.

7. EU: *European Parliament and Council Directive on Packaging and Packaging Waste*. European Parliament, Brussels; 1994.

8. EU: *Council Directive on Batteries and Accumulators Containing Certain Hazardous Substances*. European Parliament, Brussels; 1991.

9.  Jayaraman V, Guide VDR, Srivastava R: A closed-loop logistics model for remanufacturing.. *Oper. Res. Soc.* 1999, 50:497-508.

10. Hammond R, Amezquita T, and Bras B: *Issues in the automotive parts remanufacturing industry - a discussion of results from surveys performed among remanufacturers.* 1998.Accessed 19 July 201

11. Guide VDR, Muyldermans L, and Van Wassenhove LN: Hewlett-Packard company unlocks the value potential from time-sensitive returns.*Interfaces* 2005, 35:281-293.

12. Van Nunen JAEE, Zuidwijk RA: E-enabled closed-loop supply chains.*Calif. Manage. Rev.* 2004, 46:40-54.

13. Maslennikova I: Xerox's approach to sustainability. *Interfaces* 2000, 30:226-233.

14. Johnson MR, Wang MH: Planning product disassembly for material recovery opportunities. *Int. J. Prod. Res.* 1995, 33:3119.

15. Feldmann K, Meedt O, Trautner S, Scheller H, Hoffman W: The "green design advisor": a tool for design for environment. *Electron. Manuf.* 1999, 9:17.

16. Rose CM, Ishii K: Product end-of-life strategy categorization design tool. *Electron. Manuf.* 1999, 9:41.

17. Bufardi A, Gheorghe R, Kiritsis D, Xirouchakis P: Multicriteria decision-aid approach for product end-of-life alternative selection. *Int. J. Prod. Res.* 2004, 42:3139-3157.

18. Guide JV, Daniel R, Souza GC, van der Laan E: Performance of static priority rules for shared facilities in a remanufacturing shop with disassembly and reassembly. *Eur. J. Oper. Res.* 2005, 164:341-353.

19. Guide VDR: Scheduling using drum-buffer-rope in a remanufacturing environment. *Int. J. Prod. Res.* 1996, 34:1081.

20. Goldratt EM: *The goal: beating the competition.* Hounslow, Creative Output; 1986.

21. Chung C-J, Wee H-M: Short life-cycle deteriorating product remanufacturing in a green supply chain inventory control system. *Int. J. Prod. Res.* 2011, 129:195-203.

22. Richter K: Pure and mixed strategies for the EOQ repair and waste disposal problem. *OR Spektrum* 1997, 19:123-129.

23. Kondoh S, Salmi T: Strategic decision making method for sharing resources among multiple manufacturing/remanufacturing systems. *J. Remanuf.* 2011, 1:1-8. BioMed Central

24. Konstantaras I, Skouri K, Jaber MY: Lot sizing for a recoverable product with inspection and sorting. *Comput. Ind. Eng.* 2010, 58:452-462.

25. Nenes G, Panagiotidou S, Dekker R: Inventory control policies for inspection and remanufacturing of returns: a case study. *Int. J. Prod. Econ.* 2010, 125:300-312.

26. Uzsoy R: Production planning for companies with product recovery and remanufacturing capability. In *Proceedings of the 1997 IEEE International Symposium on Electronics and the Environment, 1997 (ISEE-1997), San Francisco, 5-7 May 1997.* IEEE, New York; 1997:285-290.

27. Spengler T, Schroter M: Strategic management of spare parts in closed-loop supply chains– a system dynamics approach. *Interfaces* 2003, 33:7-17.

28. van der Laan E, Salomon M, Dekker R: An investigation of lead-time effects in manufacturing/remanufacturing systems under simple PUSH and PULL control strategies. *Eur. J. Oper. Res.* 1999, 115:195-214.

29. Guide JV, Daniel R, Srivastava R: An evaluation of order release strategies in a remanufacturing environment. *Comput. Oper. Res.* 1997, 24:37-47.

30. Guide VDR, Jayaraman V, Srivastava R: The effect of lead time variation on the performance of disassembly release mechanisms. *Comput. Indust. Eng.* 1999, 36:759-779.

31. Guide VDR, Srivastava R: Inventory buffers in recoverable manufacturing. In: Proceedings of the 28th Annual Meeting of the Decision Sciences Institute Atlanta. *Georgia USA* 1997, 1–3:1405-1407.

32. Peng ZY, Zhong DY: Optimization model for closed-loop logistics network design in manufacturing and remanufacturing system. In *International Conference on Service Systems and Service Management, Chengdu, 9-11 June 2007* Edited by Zhong DY. 1-4.

33. Blackburn JD, Guide VDR Jr, Souza GC, Van Wassenhove LN: Reverse supply chains for commercial returns. *Calif. Manage. Rev.* 2004, 46:6-22.

34. Ray S, Boyaci T, Aras N: Optimal prices and trade-in rebates for durable, remanufacturable products. *Manuf. Serv. Oper. Manag.* 2005, 7:208-228.

35. Xiaochen S, Yancong Z, Yuling N, Guangwei S, AGS: Optimal Control for a Remanufacturing Reverse Logistics System under Buy-Back Policy. In *Yancong, Z ISDA '06 Sixth International Conference on Intelligent Systems Design and Applications 2006, vol. 1.* IEEE, Washington; 2006:1197-1202.

36. Guide VDR, Van Wassenhove LN: The reverse supply chain. *Harv. Bus. Rev.* 2002, 80:25-26.

37. Steinhilper R: *Remanufacturing: The Ultimate Form of Recycling.* Fraunhofer IRB, Stuttgart; 1998.

38. Goldberg LH: *Green Electronics / Green Bottom Line: Environmentally Responsible Engineering.* Woburn, Butterworth-Heineman; 2000.

39. Nagel C, Meyer P: Caught between ecology and economy: end-of-life aspects of environmentally conscious manufacturing. *Comput. Ind. Eng.* 1999, 36:781-792.

40. Brent AC, Steinhilper R: Opportunities for remanufactured electronic products from developing countries: hypotheses to characterise the perspectives of a global remanufacturing industry. In *IEEE AFRICON Conference in Africa, Gaborone. Volume 892.* Edited by Steinhilper R. IEEE, Piscataway; 2004::891-896.

41. Voss C, Tsikriktsis N, Frohlich M: Case research in operations management. *Int. J. Oper. Prod. Manag.* 2002, 22:195-219.

42. Meredith J: Building operations management theory through case and field research. *J. Oper. Manag.* 1998, 16:441-454.

43. Eisenhardt KM: Building theories from case study research. *Acad. Manage. Rev.* 1989, 14:532.

44. Ijomah WL, Childe SJ: A model of the operations concerned in remanufacture.*Int. J. Prod. Res.* 2007, 45:5857-5880.

45. Aguilar-Savén RSRS: Business process modelling: review and framework. *Int. J. Prod. Res.* 2004, 90:129-149.

46. Ang CL, Luo M, Gay RKL: Knowledge-based approach to the generation of IDEF0 models. *Comput. Integr. Manuf. Syst.* 1995, 8:279-290.

47. Loomba APS, Nakashima K: Enhancing value in reverse supply chains by sorting before product recovery. *Prod. Plan. Control* 2011, 23:205-215.

48. Depuy DW, Usher JS, Walker RL, Taylor GD: Production planning for remanufactured products. *Prod. Plan. Control* 2007, 18:573-583.

49. Deming WE: *Out of the Crisis*. MIT, Cambridge; 1982.

# Chapter 3

# Urban Public Space between Fragmentation, Control and Conflict

Alfredo Mela

Politecnico di Torino, Torino, Italy

## ABSTRACT

The article is focused on the different tendencies affecting urban public space in contemporary cities. It is based on a reflexion on some emerging themes in the recent debate in urban studies, paying particular attention to the approaches that emphasize the fragmentation of public space and the presence of control strategies, highlighting the function of tecnologies and material elements of built environment. The main thesis of the article is that public space, far from having become marginal in a context where virtual relations have a growing importance, is a field in which various types of dialectical tensions operate. In particular, at the one hand, in different contexts it is possible to recognize the presence of a complex strategy of domestication and control of urban places, linked to a process of commodification and privatisation. On the other hand many types of opposing practices and movements are also present, that propose an alternative project of use. In this framework, public space is both a place of confrontation between opposing tendencies and a stake, on which future city models depend significantly.

# INTRODUCTION

Public space is a crucial theme in the urban studies of this first part of the 21$^{st}$century for theoretical reasons and, at the same time, for its practical-political relevance. From a theoretical point of view, the transformations of public space and its crisis are often analysed as a key for the study of socio-spatial changes in post-industrial society and, in more general terms, as an interpretative factor of social relations in contemporaneity. From a political point of view, public areas of the city are examined as one of the arenas in which the contradictions and conflicts, typical of the current phase, are displayed; contradictions that take place at a macro-social level – for example between large renewal projects of urban centres and the needs of populations threatened with expulsion – as much as on the micro level, between practices of specific groups and social actors in daily life frameworks.

This article aims to focus, in fact, on this plural and often conflicting dimension of urban space, trying to show how – far from having become marginal in a society in which virtual interactions have a primary importance – it could be represented as a field in which various types of dialectical tensions operate, on which the power relations and lifestyles of contemporary societies significantly depend. In particular, on the one hand contemporary cities are affected by processes of control and normalization of citizens' behaviour; on the other hand they are witnessing the development of social practices aimed at countering the control and to propose alternative ways of use of public space. In paragraph 2 we take into account some interpretative lines in the debate of social sciences on the city, focussing on those which study tendencies of domestication and control of public places. Paragraph 3 examines the role of technology and the material dimension of built environment; in the 4$^{th}$ we reflect on the practices and bottom up movements which represent a form of resistance and an alternative to the dominating strategies of urban governance. Finally, in the conclusion we highlight the characteristics of contemporary public space as a place where a complex set of conflicting practices take place and as a stake in the construction of new urban lifestyles.

# PUBLIC SPACE, PACIFICATION AND CONTROL

A large part of the debate on public space in the contemporary city revolves around the analysis of crisis factors of public space, or rather the transformation trends of the post-industrial societies that bring a radical change in the functions of urban places, their meanings and symbolism, the practices that are carried out in them. We could say that this debate is structured on different levels. On the one hand, there are the reflections of a more abstract order, which examine the consequences on an urban scale of processes characterising contemporary societies on a global scale, such as the transformations of governance and urban policies (Mac Leod [2011]), public sphere (Castells [2008]); the processes of metropolisation, loss of city limits (Gillham [2002]), increasing mobility, growth of digital connections and on line communication, and change of space and time structures (Smith [2003]). On the other hand, there are contributions that analyse urban public spaces in specific contexts, as well as projects and practices that characterise them, often focussing on particular types of place, such as green areas and urban agriculture fields (Bergamaschi [2012]), libraries (Given et al. [2003]), town squares, shopping areas, pavements (Loukaitou-Sideris and Ehrenfeucht [2009]), sports areas (Puig et al. [2006]) and so on. In these cases, reflections often start with empirical analyses on individual cities or with comparative analysis between different contexts (Mazzette [2013]) and attempt to link the results of fieldwork to the themes of international theoretical debate.

We could observe that often more abstract reflections tend to highlight crisis factors, emphasising the split between traditional models of public space, or even of industrial modernity models, and the current postmodern forms. Instead, in many cases the analyses that derive from empirical studies highlight at the same time the crisis and the persistent vitality of urban places they study, showing the evolution of practices deriving from the emergence of new actors or urban populations, or from innovative design trends, as well as the presence of new forms of social conflict linked to the city use.

In any case, there are themes of debate that represent a sort of bridge between theoretical and empirical studies. A central topic refers to the fragmentation of the urban public space, a phenomenon also connected with the privatisation trends the city (Kohn [2004]). The analysis of these processes lends itself to an interpretation that starts from the macro-social level, as well as to an observation based on the urban micro-spaces. From a more general viewpoint, the segmentation and specialisation of the urban space is, at the same time, the spatial reflection of processes that regard the social and cultural sphere – and which lead to the multiplication of groups and lifestyles – and the effects of capitalist and neoliberal policies, which lead to the reduction of the common goods sphere and the appropriation of them by the market.

In this perspective the analyses of David Harvey have a central place; this author, throughout his work, has always underlined both the central role of processes of urbanisation in capitalist accumulation, and the importance of urban struggles as a fundamental axis of the opposition to capitalism. In his more recent works, (especially Harvey [2010], [2012]), he particularly insists on the structural character of "accumulation by dispossession", the mechanism of creative destruction through which the market takes possession of common goods. He also strongly emphasizes the urban origins of the crisis of the capitalistic model. Furthermore, he highlights that the great urban transformations of contemporary history (from that of Paris in the Second Empire, to the American suburbanisation of the second post-war period, up to the globalised processes of urbanisation of the last few decades) have always caused radical changes in lifestyles and in power relationships between social groups. In this regard, a distinctive characteristic of the current transformations is the multiplication of urban market niches, which "suffuses the contemporary urban experience with an aura of freedom of choice, provided you have the money" (Harvey [2010], p. 175).

We could say that this fragmentation of the market and consumption, in their various aspects, correspond to the subdivision of the city into spaces – or rather into space-time frames – each of which takes on a peculiar character and aims at a particular target of consumers. The theming of the urban areas and the presence of a dialectic between inclusive and exclusive mechanisms is a consequence of these trends, favoured by the sprawling form of contemporary urbanisation. The metropolitan sprawl – which, after all, shows different patterns in each context – and the diffusion of urban poles also in extra-urban territory facilitates the physical separation between specialised fragments of public space aimed at different social or consumers groups.

This phenomenon, however, presents contrasting aspects and is not free of contradictions. On the one hand, it takes on the character of pacification and domestication of places. This is the dimension that focusses more on inclusion than exclusion, or rather, which stakes more on the attraction capacity of certain public spaces for specific users, in ways that mask the repulsion aspects for other potential users. To this end, the concept of "ambient power" proposed by Allen ([2006]) is interesting, highlighting how the creation of a specific atmosphere in particular urban places makes some practices easier and discourages others. It is not solely about a *mise-en-scene*professionally combining various types of sensorial stimulation, from the experience of walking (Degen and Gillian [2012]) to the visual and olfactory (smellscapes: Henshaw [2013]) passing through the creation of soundscapes (Atkinson [2007]). It is especially important to favour a soft and domestic experience of these spaces leaving the processes of control and selection in the background, while remaining in any case constantly in operation. Shopping areas, be they malls or public roads destined to tourists and gentrifiers, are a paradigmatic example of this type of strategy based fundamentally on seduction.

On the other hand, though, are the strategies in which the function of public space control, instead of being left in the background, is clear and takes on a visibly exclusive character. Large, extraordinary events hosted by cities, such as the Olympics or other large sporting, political or religious events, are emblematic of these strategies that use complex surveillance assemblages (Boyle [2012]). The aspect of seduction and beautification (Newton [2009]) is also strongly highlighted in these situations: just think, for example, of the urban cosmetic strategies undertaken to hide the favelas in Rio de Janeiro in view of the 2014 World Cup and the 2016 Olympics, trying to commercialise a stereotyped and sweetened image of them (Steinbrink[2013]). Nonetheless, just like the cases of Rio and London Olympic Games show, strategies to boost security in the city and to eliminate - or at least to hide - all possible activities in contrast with the city image they want diffuse are put in the full light, to reassure the public of sporting events and to guarantee profit for the investors. The case of mega-events, after all, is just one of the most visible aspects of a policy of growing control on the city and its spaces, leading some author to consider it a true process of urban militarisation (Graham [2010]), justified as measures against terrorism, clandestine immigration or simply as a response to increasing worries for the safety of urban spaces. It is a process where many factors contribute, including the "ordinary" technologies of urban control and the diffusion of a media culture inspired by military violence and the development of an economy based on the industry of security (Graham [2012]).

Seduction and control strategies operate in a synergic way and create evident effects on the form of urban public space. To recall a frequently used metaphor, these effects can be described as a process of capsularisation of metropolitan areas (De Cauter [2004]): the city is fragmented into a multiplicity of closed or, in any case, controlled spaces, that defend the occupants from unwanted stimuli and which regulate their behaviour. Residential areas – in particular gated communities – and commercial ones, such as university campuses, leisure centres, tourist and cultural areas, represent "real" capsules which, in turn, interconnect with "virtual" ones that are always accessible through the internet and social networks. Daily life is carried out predominantly in capsules or along the channels of communication that join them: motorways, underground systems, metropolitan railways, cycle paths. Beyond this network are the marginal areas, in many contexts left, specially but not exclusively in Global South metropolises, to spontaneous forms of control and, often, dominated by criminal organisations, which means that these areas also become particular types of capsule. In this perspective, then, both capsules and channels of interconnection lose their image as "meeting places" and free access zones, an image which is traditionally linked to public spaces. They are privatised and lend themselves especially to mono-dimensional and individual use, paradoxically leaving a truly public function only to interstitial areas, which are not controlled by any public or private agencies.

# TECHNOLOGY AND MATERIAL DIMENSION OF THE SPACE

The hypothesis of an integral capsularisation of the city could be criticised as a dystopic vision - implying a unidirectional evolution of urbanism that requires the destruction of the public space – without offering a way out. As we will see below, however, there are processes in contemporary cities that move in the opposite direction and configure public places as a space in which to build alternative projects for the city. In any case, the capsularisation hypothesis has the advantage of clearly indicating a risk that is linked to the increasingly capillary role of urban technologies and the various interfaces between "real" and "virtual" worlds.

This evokes a very important theme of the current debate regarding the public space and on the projects that regard it: the key-word, on which this debate is focussed, is the transformation of the city into a "smart" city. In recent years, the concept of smart city has had huge success not only in the academic field but also – and above all – in administrations and institutions of different level. Despite this, its definition remains open to multiple interpretations. The broadest definitions tend to use this concept as a general paradigm of a desirable future city, including many integrating dimensions: economy, mobility, and environment, social relations, lifestyles, governance are all fields of application of the smartness philosophy (Chourabi et al. [2012]. Moreover, this vision is based on the idea that the key to success of every possible urban policy or project is in the application of innovative technologies and, in particular, of ICTs. This idea is explicit, for example, in the proposed definition in a recent study promoted by the European Parliament, in which it is said that "a Smart City is a city seeking to address public issues via ICT- based solutions on the basis of a multi - stakeholder, municipally based partnership" (European Parliament [2014], p. 17).

Concerning specifically public spaces, there are multiple applications that regard, for example, the transport system, tourist areas, commerce, museums and cultural activities. Furthermore, ICTs can ease use of all types of public space, favouring an increase in security, reducing road risks, controlling environmental parameters and so on. Taking into account this potential distribution, smart cities are often presented as the proposal of an inclusive city, whose public spaces are strongly interconnected and open to generalised use.

However, in a conspicuous part of literature and, above all, in the experiences presented as concrete applications of the smartness philosophy, there are often aspects that bring forth doubts on the effective inclusivity of smart cities (Santangelo et al. [2013]).

Firstly, it can be noted that, beyond the rhetoric of integration and multidimensionality of the concept, in many cases not only is the accent placed almost exclusively on technology, but there is also the expectation that their development would lead to an opening of new markets. In this sense, therefore, we can suspect that there is a close relationship between smart cities and the policies aimed at the idea of an entrepreneurial city (Hollands [2008]). It is because of this research into market-oriented technological solutions that we can then observe that not all urban spaces, nor all their users, are as likely to undergo innovative projects. Quite the contrary – the applications are more likely to be aimed first at areas of particular economic interest (for example, tourism, commerce, showbusiness, as well as stations, airports, gentrified central areas) and less likely to affect places in which the poorest part of the population live, unless – perhaps – to increase control over them. If this were found to be the case, it would end up seconding, instead of contrasting, the fragmentation or even the capsularisation of public spaces, highlighting the imbalance between public environments that take on a smart look and those that maintain a "non-intelligent" character.

The attention in the technological dimension in its relationship with the city and public spaces does not only regard the smart city debate. More in general, as Saskia Sassen states, today we can observe an increasingly striking interconnection between "real" and "virtual" urban spaces and even a sort of interexchange between them: "much of what is liquefied and circulates in digital networks and is marked by hypermobility, actually remains physical and hence possibly urban in some of its components. At the same time, however, that which remains physical has been transformed by the fact that that is represented by highly liquid instruments that can circulate in global markets" (Sassen [2006], p.24).

It can be noted that the attention to digital technology in current debate is often linked to an increasingly wide discovery of the materiality of the city and its public spaces. This implies an increasing interest in the physical reality of the built environment and in the ways the non-human elements of public space (the "actants", to use Latour's [2005] concept) interact with the social actors' behaviour. In many recent contributions, the influence of the Actor Network Theory (ANT) has in fact led to sociological interpretations not limited to the human sphere but able to also include in their analysis actants playing an active role in social contexts. In this, Latour himself also contributed with his polemic against a representation of the objects anchored to the Cartesian conception of the *res extensa* as an inert material, on which only the *res cogitans* actively works. This old conception contrasts with the emerging idea that the material elements of built environment - be they technological devices (Crang and Graham [2007]), architectures, urban furniture - can actively interact with human action, for example by regulating the use of a space, drawing the attention of people, concentrating and mixing the flows of actors and so on (Latour and Yaneva [2008]). This idea must not be confused with the affirmation of determinism of space on human behaviours: rather, it intends to affirm an interaction characterised by a reciprocal influence of actors and actants and their interconnection that leads to forming complex networks.

The reference to the ANT and the attention to the role of material elements could help us to understand the forms in which different ways of using the territory are stabilised and established in close relation with specific settings in the built environment. This stabilisation, in many cases, may also be considered what is at stake in conflicts rising in public areas and may become the objective of explicit strategies of control and domestication (Kärrholm [2008]). So, we need to underline the close relationship between the processes of gentrification and commercialisation, which take place in central areas of the city or in the regeneration of ex-industrial areas, and the pedestrianisation of some roads, the presence of security cameras, the creation of an urban setting that encourages people to spend time outside. We could, therefore, conclude that the attention to materiality and the activity of the built space helps us to better understand the many strategies of seduction and control, clarifying in which ways these strategies boast design that involves at the same time actors and actants, favouring reciprocal interactions and connecting them on the sensorial, functional and symbolic levels.

# CONFLICT AND PRACTICES

Until now, we have mainly highlighted the tendencies to the transformation of public urban places that come from above, or rather from strategies put in place by public or private institutions or, again, by public-private coalitions and partnerships, aimed at controlling the spaces in view of their economic valorisation and neutralising any practices that contrast with this goal. These strategies, however, are often at odds with the opposition of social actors who act in an organized way to counter them or, simply, depict alternatives for the organisation and use of urban public places through their practices and lifestyles.

Regarding these practices, it is good to note how they present a high level of heterogeneity and which – while opposing themselves to strategies of privatisation and control of urban spaces – may lead to different results. To schematise, we can identify two distinct axes, based on which it is possible to classify the bottom up practices which, overall, represent an alternative to top down strategies.

The first axis regards the grade of intentionality and structuring of said practices. On one extreme are the processes based on an alternative project of definition and use of public spaces. In this case, moreover, the practices are structured in actual projects of citizenship, which not only are expressed through behaviour but also propose explicit alternative models of public space (Tamayo[2006]): these therefore give place to social movements or, in any case, to collective forms of expression. On the other extreme, there are processes that derive spontaneously from individual or group behaviour so they propose a non-conventional use of the public space, without defining proposals or projects for transformation. These may derive, for example, from the presence of "new populations" in urban space – like new immigrants or the young protagonists of the nightlife (Hael [2011]; Mela [2014]) – or from the diffusion of new lifestyles and behavioural patterns.

The second axis regards, on the other hand, the nature of the practices and activities with which these counter the strategies of privatisation and commercialisation of the city. On one hand, we can place those practices that fundamentally express a desire to participate in decisions regarding public space and its management: people acting in this way, therefore, consider (explicitly or implicitly) public space as common property that must be defended and conserved to keep it open to multiple kinds of use and types of users. Often these actors are willing to collaborate with public institutions, as long as these are on the same page and are open to citizens' participation.

On the other hand, we find the practices that have an essentially antagonistic meaning: some groups or sectors of the urban population are opposed to projects or policies carried on by public institutions. These actors refuse (explicitly or implicitly) each form of co-participation or dialogue with institutions or with differently characterised groups, emphasizing their identity and autonomy.

Figure 1 shows how the intersection of these axes ends up structuring a field of practices put into act by individuals or groups and subdivided this field into four quadrants.

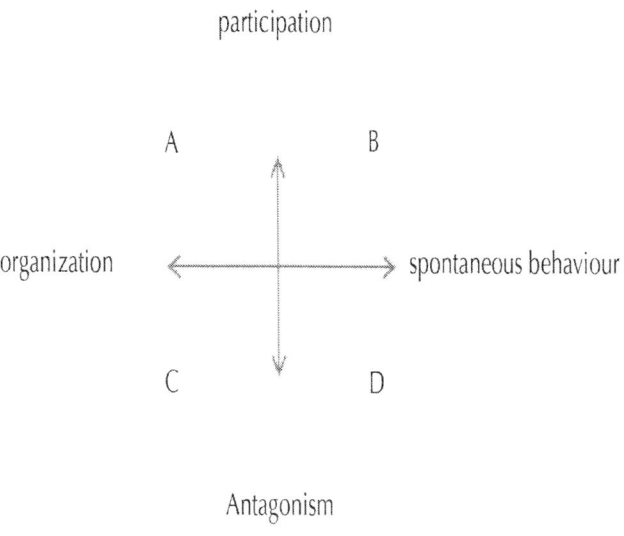

**Figure 1**: Bottom up practices participation.

In quadrant A, therefore, are the collective movements of organised citizens and civil society associations, that aim to have more weight in planning and managing urban public space. They act in a participatory way while also expressing their disagreement when it is useful for the success of their action. In quadrant B, we find alternative uses of public space which display merely an active and intense use of it, in contexts in which – to repeat Amin's concepts ([2008]) – the ethics of the situation, influenced by the physical configuration of places in a context of throwntogetherness (Massey [2005]), favour practices that ensure a variety of uses and accept diversity. In quadrant C, there are different types of antagonist organised movements that use public space as a place to conflict with power, in forms that consider the control of specific spaces as a decisive action in practical and symbolic terms. Finally, in D, are non-organised behavioural forms which, nonetheless, have the effect of favouring the control of public spaces by marginal groups or by actors with alternative lifestyles.

It is obvious, after all, that this classification of the practices must not be intended as a rigid division in clearly distinct categories. Often the phenomena that we truly observe in urban public spaces are mixed, and we see that even the same phenomenon can take on different connotations in time and space, depending on context variables and on its evolution for internal dynamics or in reply to external interventions.

# PUBLIC SPACE AS A PLACE OF DIALECTIC TENSIONS

The reflections in this article have highlighted, above all, the great complexity of the tensions to which public spaces in contemporary cities are subject and the presence of opposing strategies and practices aimed to their organisation and use.

In summary, we can say that today, in most urban contexts, we are witnessing the intensification of strategies that imply the fragmentation of the public space – that is, the increase of zoning practices that do not correspond to a public plan but more to demands of the market –urban sprawl and the commercialisation of places (Borja [2003]). These strategies present two different – seemingly opposing - versions that are, in reality, complementary: one that aims at seducing the various types of consumers and tries to domesticate urban places, and one that is expressed through explicit control and exclusion. The first version, the soft one, creates comfortable environments that are configured like capsules adapted to specific forms of consumption; the second, the hard one, guarantees a security of use by expelling every form of behaviour and, even, the physical presence of social groups incompatible with the needs of private economy and political power. Technology often has an essential role in both versions - beyond the rhetoric that exalts their smartness – as do the physical elements of the built space, highlighting their active role in the interaction with the behaviours of social actors.

Strategies of this nature have a top-down character, or rather they are the result of a process of governance of the city and its public spaces that starts from a network of public and private actors. Nevertheless, these are effective only when they actually find a positive response from wide social groups or are able to prevail over opposing tendencies. Therefore this effectiveness is not always guaranteed; in fact, even the public space of contemporary cities – though in a different form than in the past – is a place of resistance against the dominating strategies, which are bottom up. These opposing processes are not always well organised, nor do they necessarily express conscious alternatives: sometimes they occur purely as practices in public spaces that actually conflict with those proposed by the dominant models of governance. In different situations they may express a genuine aspiration to an inclusive participation or themselves act as forms of occupation and control of the urban spaces with exclusionary effects in relation to other actors or other practices.

In any case, the presence of these alternatives, even with the ambivalence that distinguishes them, highlight the fact that urban public space is still at the centre of dialectic tensions. We cannot predict how these tensions will end, though they are of great importance in creating relationships of power among social groups and in defining the very meaning of cohabiting in the city. Furthermore, this shows how the form of urban space, in its physical and symbolic dimension, is not only a question of aesthetic or functional choices, but is a decisive factor in the dialectics between individual and collective social actors and plays an important role in the prevalence of inclusive or exclusive models of social relations.

# REFERENCES

1.   Allen J (2006) Ambient power: Berlin's Potsdamer Platz and the seductive logic of public spaces. Urban Stud 43(2):441-455

2.    Amin A (2008) Collective culture and urban public space. City 12(1):5-24

3.    Atkinson R (2007) Ecology of sound: the sonic order of urban space. Urban Stud 44(10):1905-1917

4.    Bergamaschi M (2012) Coltivare in città. Orti e giardini condivisi. Sociol Urbana e Rural 98:7-11

5.    Borja J (2003) La ciudad conquistada. Alianza Editorial, Madrid.

6.    Boyle P (2012) Securing the Olympic Games: Exemplifications of Global Governance. In: Lenskyj HJ, Wagg S (eds) The Palgrave Handbook of Olympic Studies, Palgrave MacMillan, Houndmills, Basingstoke.

7.    Castells M (2008) The New Public Sphere: Global Civil Society, Communication Networks, and Global Governance. Ann Am Acad Pol Soc Sci 2008(616):78-93

8.    Chourabi H, Gil-Garcia JR, Mellouli S, Nahon K, Nam T, Pardo TA, Scholl HJ, Walker S (2012) Understanding Smart Cities: An Integrative Framework. IEEE Computer Society, Maui.

9.    Crang M, Graham S (2007) Sentient cities: ambient intelligence and the politics of urban space. Inf Commun Soc 10(6):789-817

10.    De Cauter L (2004) Capsular Civilization: On the City in an Age of Fear. Nai Publishers, Rotterdam.

11.    Degen MM, Gillian R (2012) The sensory experiencing of urban design: the role of walking and perceptual memory. Urban Stud 49(15):3271-3287

12.    (2014) Mapping Smart Cities in the EU.

13.    Gillham O (2002) The Limitless City. A Primer on the Urban Sprawl Debate. Island Press, Washington, D.C.

14.    Given LM, Gloria J, Leckie GJ (2003) "Sweeping" the library: mapping the social activity space of the public library. Libr Inf Sci Res 25(2003):365-385

15.    Graham S (2010) Cities under Siege. The New Military Urbanism, Verso, London.

16.    Graham S (2012) When life itself is war: on the urbanization of military and security doctrine. Int J Urban Reg Res 36(1):136-155

17.    Hael L (2011) Dilemmas of the nightlife fix: post-industrialization and the gentrification of nightlife in New York City. Urban Stud 48(6):3449-3465

18. Harvey D (2010) The Enigma of Capital and the Crises of Capitalism. Profile Books, London.

19. Harvey D (2012) Rebel Cities. From the Right to the City to the Urban Revolution. Verso, London-New York.

20. Henshaw V (2013) Smellscapes. Understanding and Designing City Smell Environments. Routledge, London-New York.

21. Hollands RG (2008) Will the real smart city please stand up? Intelligent, progressive or entrepreneurial? City 12(3):303-320

22. Kärrholm M (2008) The Territorialization of a Pedestrian Precinct in Malmö: Materialities in the Commercialisation of Public Space". Urban Stud 9(45):1903-1924

23. Kohn M (2004) Brave New Neighborhoods. The Privatization of Public Space. Routledge, London-New York.

24. Latour B (2005) Reassembling the Social: An Introduction to Actor Network Theory. Oxford University Press, Oxford.

25. Latour B, Yaneva A (2008) Give me a Gun and I will Make All Buildings Move. In: Geiser R (ed) Explorations in Architecture: Teaching, Design, Research, Birkhäuser, Basel. pp 80-89

26. Loukaitou-Sideris A, Ehrenfeucht R (2009) Sidewalks. Conflict and Negotiation over Public Space. The M.I.T. Press, Cambridge (Ma) – London.

27. Mac Leod G (2011) Urban politics reconsidered: growth machine to post-democratic city? Urban Stud 48(12):2629-2660

28. Massey D (2005) For Space. Sage, London.

29. (2013) Pratiche sociali di città pubblica. Laterza, Roma-Bari.

30. (2014) La città con-divisa. Angeli, Milano.

31. Newton C (2009) The reverse side of the medal: about the 2010 World Cup and the beautification of the N2 in Cape Town. Urban Forum 20(1):93-108

32. Puig N, Vilanova A, Camino X, Maza G, Pasarello M, Juan D, Tarragó R (2006) Los espacios públicos urbanos y el deporte como generadores de redes sociales. El caso de la ciudad de Barcelona. Apunts Educ Fís y Deportes 84:76-87

33. (2013) Smart City. Ibridazioni, innovazioni e inerzie nelle città contemporanee. Carocci, Roma.

34. Sassen S (2006) Public interventions: the shifting meaning of the urban condition. Open 11:18-27

35. Smith RG (2003) World city topologies. Prog Human Geogr 27(5):561-582

36. Steinbrink M (2013) Festi*favel*isation: mega-events, slums and strategic city-staging – the example of Rio de Janeiro. Die Erde 144(2):129-145

37. Tamayo S (2006) Espacios de ciudadanía, espacios de conflicto. Sociológica 21(61):11-40

# 4

# How Buyers Perceive the Credibility of Advisors in Online Marketplace: Review Balance, Review Count and Misattribution

Kewen Wu, Zeinab Noorian, Julita Vassileva, and
Ifeoma Adaji

Department of Computer Science, University of Saskatchewan, Saskatoon, Canada

## ABSTRACT

In an online marketplace, buyers rely heavily on reviews posted by previous buyers (referred to as advisors). The advisor's credibility determines the persuasiveness of reviews. Much work has addressed the evaluation of advisors' credibility based on their static profile information, but little attention has been paid to the effect of the information about the history of advisors' reviews. We conducted three sub-studies to evaluate how the advisors' review balance (proportion of positive reviews) affects the buyer's judgement of advisor's credibility (e.g., trustworthiness, expertise). The result of study 1 shows that advisors with mixed positive and negative reviews are perceived to be more trustworthy, and those with extremely positive or negative review balance are perceived to be less trustworthy. Moreover, the perceived expertise of the advisor increases as the review balance turns from positive to negative; yet buyers perceive advisors with extremely negative review balance as low in expertise. Study 2 finds that buyers might be more inclined to misattribute low trustworthiness to low expertise when they are processing high number of reviews. Finally, study 3 explains the misattribution phenomenon and suggests that perceived expertise has close relationship with affective trust. Both theoretical and practical implications are discussed.

# INTRODUCTION

In an online marketplace, buyers rely heavily on reviews posted by advisors. A recent business survey reported that 92% of online consumers read advisors' reviews before they make purchase decisions [1]. Literature also suggests that advisors' reviews significantly influence consumers' attitudes towards the products or sellers, which ultimately influence sales [2], [3].

The extent to which a buyer accepts or follows an opinion presented in a review is a matter of persuasiveness. The persuasiveness of an online review is determined by the credibility of its source (the advisor), because online reviews are written by advisors with varied backgrounds and motivations [4]. Advisors can write reviews no matter if they are capable of assessing a product critically or not (e.g., layperson versus expert). Moreover, many intentional and unintentional factors can influence the writing of a review [5]-[7]. For instance, an advisor's account may be controlled by a seller to write positive reviews and promote himself (known as ballot stuffing); and it may also be controlled to write negative reviews to attack competitors (known as bad-mouthing). These reputation manipulation activities have been identified as a pervasive phenomenon in online marketplaces [5],[8]. Even if an advisor is a real buyer, he may still be influenced by others and write reviews that do not represent his actual experience (e.g., herd effect).

Given the uncertainty regarding the source of online reviews, buyers are motivated to assess the credibility of advisors based on accessible pieces of information [9]. Many online marketplaces (e.g., Amazon, Taobao) allow buyers to visit advisors' profile page. To evaluate an advisor's credibility, buyers are inclined to seek and use profile information as cues, other than the review itself. A number of studies have been conducted to evaluate how advisors' profile influences buyers' perception of credibility [10],[11]. Advisors' static profile information, such as real name, location, nickname and hobbies, have been found to be helpful in supporting consumers' judgment[11],[12]. However, current studies on advisors' review history mainly come from computer science field, and little is known about the impact of advisors' review history on buyers' perception of advisors' credibility. Analyzing an advisor's review history could provide useful information (e.g., purchase frequency, areas of interests or even background) about the advisor, which can be helpful for buyers to make judgement on advisors' credibility.

In this paper, we segment advisors into five types based on the ratio of positive to negative reviews (referred to as review balance). If the proportion of positive (negative) reviews is extremely higher than, substantially higher than, or almost equal to the proportion of negative (positive) reviews, the review balance is respectively defined as extreme positive (negative), positive (negative), or neutral. We choose review balance as representative of review history because it can be easily noticed by buyers through direct scanning of an advisor's review history list or a summary table provided by the platform. Prior studies indicate that buyers usually do not scrutinize reviews [13],[14]; they form attitude only based on the information they gain easily. Intuition also suggests that it is unrealistic for a buyer to conduct a comprehensive evaluation of review history for each advisor in the product page.

We conducted three sub-studies to explore how different review balances signal different meanings to buyers regarding the advisors' trustworthiness and expertise (two dimensions of credibility). Study 1 aims to gain a preliminary knowledge about buyers' perception of advisor's trustworthiness and expertise. Study 2 extends study 1 by using larger sample size and considering more variables. Finally, study 3 is conducted to further explain the results of previous two sub-studies.

# Research Background

## *Source Credibility: Trustworthiness and Expertise*

The concept of source credibility has received much attention from various fields related to communication, such as politics, human-computer interaction, marketing and information system. It is a multifaceted term suggesting that the positive characteristics of a message source can enhance the perceived value of message information, and thus increase the persuasiveness of the message [15],[16]. Expertise, trustworthiness and attractiveness are commonly reported as three dimensions of source credibility [17]. In this study, we considered source credibility as a two-dimensional construct, since expertise and trustworthiness are more relevant to online review context [18]. Trustworthiness describes the receiver's confidence in a source's objectivity and honesty in providing information [15]. There is a wide consensus on the positive relationship between trustworthiness and source credibility [19].

Expertise refers to a source's capability of providing correct and valid information [15]. Such capability can be technical-oriented or practical-oriented [20]. Technical expertise reflects the skillfulness of processing special knowledge required by writing comments towards a given product (e.g., an advisor who majors in acoustics writes a review about a headphone). Practical expertise is the skills that are gained from direct participation in related activities (e.g., an advisor who has tried many headphones writes a review about one headphone). The characteristics of online communication (e.g., limited availability of personal information) make it difficult to identify whether an advisor is an expert or not. As a result, in online context, different results have been found regarding the relationship between expertise and source credibility. For example, some studies found that expert endorsers can lead to higher source credibility than laypersons; others found that layperson can induce higher credibility than experts; yet others found that the levels of expertise make no difference in determining the perceived source credibility [19],[21].

The complex findings on expertise imply that other dimensions of source credibility might disturb the effects of expertise. As mentioned earlier, attractiveness is not relevant to online review context. Here we only take trustworthiness as an example. On one hand, high expertise can lead to increased trust because assessments of expertise and trust both employ an attribute evaluation of trustee's identifiable actions [22]. For example, a seller's expertise reflects a buyer's identification of competencies associated with the transaction. On the other hand, as suggested by the attribution theory [23], people attribute a review to both stimulus and non-stimulus causes. When the consumer suspects that the review is not drawn based on product performance (stimulus) but on the advisors' unknown intentions (non-stimulus), they will discredit the review message. In some cases, a source may be perceived to be high in expertise but low in trustworthiness [24]. For example, people trust an expert because they think expert statements are true; however, if this expert's motivation to share is reasonably suspected, people's perception of this expert's trustworthiness will decrease. The contradictory effects (e.g., high on expertise but low on trustworthiness) may cancel each other out [25].

The above mentioned two circumstances only address the impacts of expertise on trustworthiness, that high expertise can lead to both high trust (because of belief in competency) and low trust (because of suspicious motivation). However, little is known about how trustworthiness affects expertise.

## Advisors Profile and Credibility

Previous work on credibility of online reviews can be divided into two streams. The first stream of work focuses on review itself; studies have addressed many factors such as sequence of reviews[26],[27], valence [26], volume [28], information depth [29], attribution (e.g., experience issue or product issue) [26],[27]. However, these studies generally assume reviews come from credible sources.

The second stream of work deals with the credibility of advisors. Much work has been done on evaluating the effects of advisors' profile. In real online review systems, a profile usually includes an advisor's identity-related information and review history. Advisors' identity-related information, such as real name, gender, location, nickname, hobbies and reputation (e.g., special badges such as top 50 reviewers), has been proven to be helpful for buyers' judgment [11], [12],[10]. However, limited attention has been paid on the effects of review history.

The social exchange theory suggests that people develop trust based on behavioral characteristics observed from direct experiences with the trustee [30]. The history of experience facilitates the accumulation of knowledge and thus increases the validity of knowledge-based attribution [31]. Compared to static characteristics (e.g., gender, location), buyers are able to make rational credibility judgment as they obtain greater knowledge from the review history.

Positive or negative reviews could signal different meanings to buyers, for instance, a reviewer who gives negative feedback might be perceived to be high in expertise [32]. However, few studies have considered how buyers perceive expertise from advisor's review history (e.g., review balance). Moreover, current studies on the perception of trustworthiness from advisors' review history mainly come from computer science area. The basic assumptions regarding trustworthiness and advisors' review behavior are based on three points: (1) Similarity. According to social identity theory [33], a buyer may categorize an advisor who has similar purchase history and review opinions into the same social group, resulting in increased trust towards this advisor [34],[35]. (2) Social consensus, that if an advisor holds the same opinions with the majority of advisors, his/her review is perceived as correct and would be accepted [36]. (3) Social network, that dishonest advisors (e.g., fake buyers' accounts), may share the same review behavioral pattern [37]. Given the fact that related human studies are scarce, this paper evaluates buyers' perception of advisors' credibility based on review history.

## Data Source

The review dataset used in this paper is built upon Taobao review data. We selected Taobao as our target online marketplace based on two reasons. First, Chinese online marketplaces have been growing rapidly in recent years. Taobao is the leading platform with about 90% market share. Its transaction volume is estimated to have more sales than Amazon and eBay combined in 2013[38]. Taobao is well known among Chinese communities (half a billion registered users) and it is usually considered as a typical e-commerce sample in previous studies [39]. Second, despite the huge number of transactions, Chinese online marketplaces face serious reputation manipulation problem [5]. For example, some critics estimate that about 80% of *Taobao* sellers have committed reputation manipulation activities during their businesses [40]. And it has been reported that over 1000 active trust fraud companies provide services to help sellers increase reputation and whitewash negative feedback [5]. But a recent official report shows that more than 70% online buyers choose Taobao as their primary choice [41]. Therefore, the high transaction volume, serious trust issue and being buyers' primary choice jointly make Taobao a valuable target to investigate.

We use a self-developed crawler to download real review data from Taobao during 2014-04-01 and 2014-4-20. This dataset includes the latest 180-day detailed review information about 24,287 sellers and 1,686,870 advisors who are willing to show their profile. The average number of reviews per advisor in our dataset is 116.

To prepare the dataset for our experiment, we invited four master's students to select 200 positive and 200 negative reviews from our Taobao review database. The selection of reviews was based on two criteria: (1) previous studies have shown that the different review targets (product and service) have different impacts on consumer's decision-making process [26]. Therefore, we decided to only consider product attribute-based reviews to serve as data source in our experiment. Service-based reviews were excluded because service quality is usually unstable across different buyers (e.g., delivery service might be excellent in some areas but much worse in other areas) and buyers' perception of service quality contains many subjective factors. (2) We set the length of each review to be around 30 Chinese characters (about 60 English characters), and the reasons described in each review should be clear. We built advisors' profiles based on five types of review balances (See Table 1). In the following experiment, we did not set the ratio between number of positive ratings (R) and number of negative ratings (S) close to threshold values (e.g., 0.2 for Type I), because we wanted to make different types of review balance distinguishable. For example, we set the ratio of a Type I advisor's R/S to 0.05, rather than 0.19.

**Table 1**: Five types of advisors based on different review balance

| Type | Description |
|---|---|
| I. Extremely negative Balanced | R<<S[a]: number of positive ratings are significantly lower than number of negative ratings (R/S<0.2[b]) |
| II. Negative balanced | R<S: number of positive ratings are lower than number of negative ratings (0.2≤R/S<0.7) |
| III. Neutral balanced | R≈S: number of positive ratings are approximately the same as number of negative ratings (0.7≤(R/S or S/R)≤1) |
| IV. Positive balanced | R>S: number of positive ratings are larger than number of negative ratings (0.2≤S/R<0.7) |
| V.Extremely positive Balanced | R>>S: number of positive ratings are significantly larger than number of negative ratings (S/R<0.2) |

Note: [a]: R refers to number of positive ratings/reviews; S refers to number of negative ratings/reviews; b: this ratio is only used to describe a phenomenon (e.g., R<<S) and used to manipulate of advisors' profiles. It is not a strict classification of advisors.

Wu et al.

Wu et al. Journal of Trust Management 2015 2:2, doi:10.1186/s40493-015-0013-5

# Study 1

Study 1 was designed to gain a preliminary knowledge about buyers' perception of advisor's source credibility regarding different review balances.

## *Hypotheses*

Previous studies suggest that the proportion of positive reviews is much higher than negative reviews in online review systems [42],[43]. People are reluctant to give negative feedback unless they encounter terrible experience [44]. A content analysis of eBay comments shows that 72.5% of negative reviews were related to unsatisfactory product and service, while the other 27.5% were related to sellers' attempts to exploit buyers [43]. This result suggests that terrible experience (negative feedback) usually happens due to the poor product or service quality that cannot meet buyer's expectation.

The reviewers who give negative feedback are perceived as brighter and more intelligent than those who give positive feedback [32]. They give negative reviews because they have enough knowledge to identify product issues. For instance, as a domain expert, an acoustics enthusiast gives negative feedback to a headphone due to its poor performance, while non-experts could not notice the pros and cons of this headphone. In this view, an advisor with a negative review balance might be perceived as a strict expert who is hard to be satisfied. Therefore, we hypothesize that:

H1: The level of perceived expertise of an advisor increases as the review balance changes from extremely positive to extremely negative.

Negative feedback usually contains distinctive information than positive ones, therefore, it is perceived to be more accurate, trustworthy and helpful for buyers to make decisions [42]. Absence of negative feedback may have nothing to do with the judgment of review authenticity[19]. An advisor who has almost all positive feedback (review balance: extreme positive) may be considered as a malicious account controlled by a dishonest seller to do self-promotion, or as a "Mr. Goody-goody" who always gives positive feedback regardless of his actual experience. Similarly, an advisor who gives all negative feedback (review balance: extreme negative) may be judged to be a malicious account used to attack competitors, since the case that a buyer always experiences unsatisfactory transactions is unrealistic. Previous studies have found that buyers are more likely to form positive attitudes (e.g., trust, purchase intention) towards a product which receives a mix of positive and negative reviews [45],[46],[19]. Therefore, it is reasonable to assume that an advisor who posts both positive reviews and negative reviews would be perceived as trustworthy. We hypothesize that:

H2: The level of perceived trustworthiness is high when an advisor's review balance is neutral, and the level of perceived trustworthiness is low when an advisor's review balance is either extremely positive or extremely negative. Especially, an advisor with extreme negative review balance is perceived to be most untrustworthy.

## *Experiment and Result*

In order to reduce cognitive load, we only considered ratings in this sub-study. We created two sets of advisors' profiles based on our review dataset. Advisors in each set have entirely different review balances (see Table 2). Although these advisors' profiles cannot present the characteristics of the whole dataset, using a small amount of typical experiment material is acceptable in many studies [9],[47].

**Table 2**: Advisors' profile used in study 1

| Type | Description | Set 1 (R,S) | Set 2 (R,S) |
|------|-------------|-------------|-------------|
| I | $R < < S^a$ | (5, 86), (0, 103) | (2, 42) ,(0, 63) |
| II | $R < S$ | (31, 57), (38, 64) | (13, 30) |
| III | $R \approx S$ | (51, 43), (58, 42) | (29, 24), (43, 32) |
| IV | $R > S$ | (68, 31), (72, 23) | (37, 13), (64, 14), (56, 16) |
| V | $R > > S$ | (104, 0), (115, 1) | (49, 1), (43, 1) |

Note [a]: R refers to number of positive ratings/reviews; S refers to number of negative ratings/reviews.

Wu et al.

Wu et al. Journal of Trust Management 2015 2:2, doi: 10.1186/s40493-015-0013-5

Twenty experienced online buyers were invited to evaluate the impacts of review balance on perceived trustworthiness and expertise. These participants were all aware of unfair rating/review phenomenon in online marketplaces, they were told that the rating history of each advisor in this survey was based on real data gained from Taobao. The interface of the experiment system is shown in Figure 1.

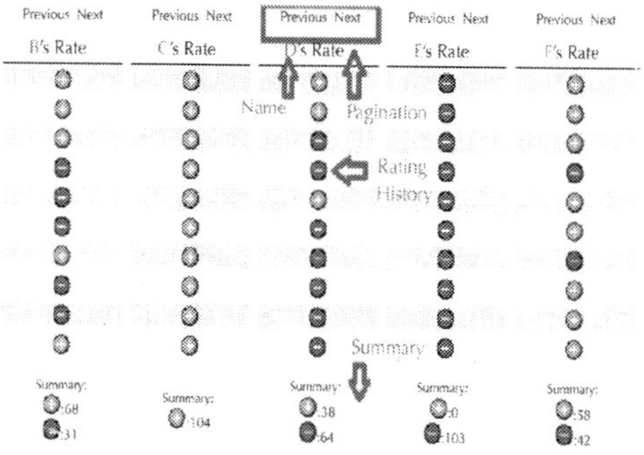

**Figure 1**: The interface of user experiment in study 1.

For the judgement of perceived trustworthiness, we randomly assigned 10 participants to check the rating history of advisors in Set 1 and asked them to rank advisors based on their perceived trustworthiness from the lowest (1) to the highest (10) on a ten-point scale (we used a computer program to ensure that each ranking position has only one advisor). Then we assigned the remaining 10 participants to rate advisors in Set 2 and rank advisors in the same way.

For the judgement of perceived expertise, we used the same advisors' profiles and the same subjects (however, two of them quitted). We randomly assigned 9 participants to check advisors in Set 1 and asked them to rank advisors based on perceived expertise from the highest to the lowest on the ten-point scale (1 shows the least expertise and 10 shows the highest expertise). Then we assigned the remaining 9 participants to check Set 2 and rank advisors, respectively.

We used Kendall's coefficient of concordance (W) to measure the degree of agreement among participants with the rankings of advisors. The capability of W in performing multiple judgments (more than two) makes it the most suitable tools to test inter-judge reliability [48]. Past studies suggest that the value of $W > 0.7$ shows strong consensus; $W = 0.5$ shows moderate consensus; and $W < 0.3$ shows weak consensus amongst different users on their ranked data [48].

In the test regarding perceived trustworthiness, for Set 1 we achieved $W = 0.7578$ ($p < 0.0001$), and for Set 2 we achieve $W = 0.7345$ ($p < 0.0001$). Therefore, there is a strong consensus between participants in terms of ranking different groups of advisors. The average ranking result shown in Figure 2 suggests that the relationship between review balances (from extremely negative to extremely positive) and perceived trustworthiness follows an inverted-U shape, and an extremely negative balanced review history is perceived as the most untrustworthy profile by buyers (2 versus 3.4 and 2.3 versus 2.95).

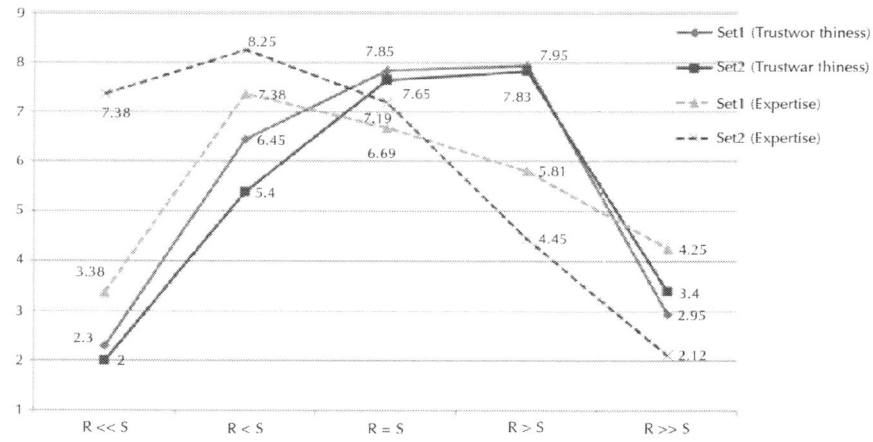

**Figure 2:** Perceived trustworthiness and expertise of advisors in study 1.

In the test regarding perceived expertise, for Set 1 we achieved W = 0.2867 (p<0.05), and for Set 2 we achieve W = 0.6451 (p<0.0001). This result indicates that the levels of consensus in Set 1 and Set 2 are weak and moderate, respectively. The averaged ranking result is shown in Figure 2, which suggests that perceived expertise does not increase linearly when review balance ranged from extremely positive to extremely negative. Meanwhile, participants' rankings about advisors with almost all negative reviews (Type I) are different (7.38 versus 3.38) across two sets.

In summary, the results from study 1 reject H1 because advisors with extremely negative review balance (Type I) were perceived to be low in expertise. H2 is supported, suggesting that advisors who always give the same ratings (either negative or positive) are not trustworthy to buyers.

Considering that the participants did not gain high consensus regarding the expertise of the advisors, it is interesting to further explore the influences of review balances on perceived credibility (especially expertise) of advisors.

## Study 2

There are at least four issues in study 1, which limit the explanation power of the result. First, the sample is relatively small (20 participants).

Second, the list of reviews only contains ratings, and it is not clear what the results would be when both ratings and comments are displayed (a real online review system usually displays both ratings and comments). Third, the measurements of trustworthiness and expertise are based on ranking, not on pre-validated questions. Ranking has its limitations, for example, it uses a one-to-one matching method between an advisor and a position and therefore, it might be difficult for participants to choose between two or more advisors when their trustworthiness/expertise perceived to be similar. Moreover, rankings only provide sequential data within a set but little is known about the differences across two sets. And fourth, the total number of reviews is not controlled.

The aim of study 2 is to further verify the results of study 1 by considering the limitations of study 1. First, a large sample was organized, including 200 participants; second, both ratings and review comments were displayed to participants; third, pre-validated questions and Likert scale were used to measure participants' opinions. And fourth, perceived trustworthiness and expertise were evaluated in both high and low review count conditions.

## Experiment Preparation

To determine appropriate number of reviews in two conditions (high and low number of reviews), we manipulated five lists of advisors' review history, which contained 10, 40, 80, 120 and 200 reviews. We provided these review history lists to three Ph.D. students who were experienced online buyers. Their feedback suggested that 10 and 40 reviews could be treated as low number of reviews, but a list with only 10 reviews was usually not enough to form an attitude towards an advisor. Therefore, we set the value of low review number to 40. The feedback also suggested that a list with 200 reviews was beyond normal processing capacity, so we set the value of high review number to 200.

We built 10 advisors' review history lists based on selected 400 reviews. The details are shown in Table 3. We edited some of the reviews to make sure that these reviews did not conflict with each other. For example, one review may indicate that an advisor is a mother, but another review may indicate that the advisor is a father.

**Table 3**: Advisors' profile used in study 2

| Type | Low review count (R,S) | High review count (R,S) |
|---|---|---|
| R<<S [a] | (1, 39) | (4, 196) |
| R<S | (12, 28) | (59, 141) |
| R≈S | (19, 21) | (98, 102) |
| R>S | (29, 11) | (136, 64) |
| R>>S | (40, 0) | (198 , 2) |

Note [a]: R refers to number of positive ratings/reviews; S refers to number of negative ratings/reviews.

Wu et al.

Wu et al. Journal of Trust Management 2015 2:2, doi:10.1186/s40493-015-0013-5

## *Details of Experiment*

We designed an online survey system which consisted of two parts: an advisor's review history and questions regarding trustworthiness and expertise. In the review history page, participants were told to imagine that they were shopping in Taobao as usual, and need to evaluate the credibility of an advisor. They should use the same amount of time to judge the advisor in our survey as in their regular purchase, and they could go to the questionnaire page as soon as they felt they have finished their judgment.

All questions in the survey were measured with 7-point Likert scale. Trustworthiness was measured by five items (dependable, honest, reliable, sincere and trustworthy); expertise was also measured by five items (expert, experienced, knowledgeable, qualified, skilled). These items were originally developed by Ohanian [25], and they have been adopted by many studies [49]. In order to do manipulation check, we used a question to ask participants to select one of the five conditions (R<<S; R<S; R≈S; R>S; R>>S) which best fits what they see.

We invited 200 participants into our experiment. They were undergraduate students and they all had purchase experience in Taobao. Each participant was randomly assigned into one of the ten conditions

(5 types of review balance×2 types of review count). Therefore, each condition had 20 participants. This sample size provided an acceptable level of statistical power with an effective size of 0.50 at a two-tailed 5% significance level [50]. We selected undergraduate students as research subjects based on following two reasons: first, students provided an accessible sample when an experiment requires a large sample size [51]; second, young adults and university students are a typical group of online buyers, and similar sampling approach has also been employed in previous studies [52],[51],[17]. Moreover, a recent official survey shows that 56.4% of Chinese buyers in online marketplaces are aged between 20 and 29, 35.9% of consumers have (or are pursuing) bachelor degrees [41].

# Analysis and Result

All participants could correctly select the condition they were assigned to, indicating that our manipulations were successful. Table 4 shows the results of factor analysis (CFA) for both high and low review count conditions. All factor loadings were significant ($p < 0.01$), and ranged from 0.73 to 0.93. The composite reliability and Cronbach's alpha of each factor ranged from 0.86 to 0.94, demonstrating acceptable levels for internal reliability (the recommended threshold for these two indices is 0.7). All values of AVE shown in Table 4 are greater than the recommended value (0.5), suggesting that the latent constructs account for the majority of the variance in their indicators on average [53]. As a common rule, the presence of multi-collinearity issue is confirmed if Variance Inflation Factor (VIF) is higher than 10 [54]. More strictly, the VIF threshold of 3.3 has been recommended by Cenfetelli & Bassellier [55]. Table 4 shows that only two items (EXP2 and EXP3) from the high number reviews group are larger than 3.3 (but smaller than 10), indicating that multi-collinearity is not a serious issue.

**Table 4:** Results from confirmation factor analysis in study 2

| Constructs | | Loading | C.R. | C.A. | AVE | VIF |
|---|---|---|---|---|---|---|
| Trustworthiness | TRU1 | 0.83/ 0.87[a] | 0.91/0.93 | 0.87/0.91 | 0.66/0.74 | 2.19/2.65 |
| | TRU2 | 0.77/0.83 | | | | 1.80/2.10 |
| | TRU3 | 0.75/0.89 | | | | 1.74/2.99 |
| | TRU4 | 0.83/0.87 | | | | 2.13/3.09 |
| | TRU5 | 0.87/0.85 | | | | 2.41/2.59 |
| Expertise | EXP1 | 0.77/0.79 | 0.90/0.94 | 0.86/0.92 | 0.65/0.76 | 1.79/2.53 |
| | EXP2 | 0.87/0.93 | | | | 2.56/4.42 |
| | EXP3 | 0.86/0.92 | | | | 2.57/3.83 |
| | EXP4 | 0.73/0.91 | | | | 1.80/2.47 |
| | EXP5 | 0.78/0.78 | | | | 1.94/2.15 |

Note [a]: the value on the left side of "/" is from the low number of reviews condition; the value on the right side of "/" is from the high number of reviews condition.

Wu et al.

Wu et al. Journal of Trust Management 2015 2:2, doi:10.1186/s40493-015-0013-5

We conducted two $5 \times 2$ ANOVA analyses on trustworthiness and expertise respectively. For trustworthiness, both review count ($F(1,190) = 4.045$, $p < 0.05$) and review balance conditions ($F(4,190) = 109.159$, $p < 0.001$) have significant main effects, but no significant interaction effect ($F(4,190) = 1.231$, $p > 0.05$). This result suggests that in general the participants perceived higher trustworthiness under the high review count conditions than under low review count conditions (mean differences $= 0.178$, $p < 0.05$). And in both low and high review count conditions, the values of perceived trustworthiness are distributed in an inverted-U curve (see the repeated contrast of means shown in Table 5).

**Table 5**: Means and repeated contrast results in study 2

| Review balance | Perceived trustworthiness | | | | Perceived expertise | | | |
|---|---|---|---|---|---|---|---|---|
| Condition | Low count [a] | Repeated contrast [b] | High count | Repeated contrast | Low count | Repeated contrast | High count | Repeated contrast |
| R<<S | 3.49 (0.72) | – | 3.42 (1.06) | – | 5.07 (0.56) | – | 3.04 (0.73) | – |
| R<S | 4.82 (0.66) | -1.33*** | 5.10 (0.54) | -1.68*** | 5.43 (0.57) | -0.36[N.S.] | 5.68 (0.73) | -2.64*** |
| R≈S | 5.37 (0.54) | -0.55* | 5.79 (0.50) | -0.69* | 4.58 (1.05) | 0.85* | 4.70 (1.07) | 0.98** |
| R>S | 6.09 (0.39) | -0.72** | 6.39 (0.39) | -0.60* | 4.40 (0.92) | 0.18[N.S.] | 4.61 (1.01) | 0.09[N.S.] |
| R>>S | 4.90 (0.57) | 1.19*** | 4.86 (0.60) | 1.53*** | 3.34 (0.81) | 1.06** | 3.35 (0.89) | 1.26*** |

Note: ***: $p < 0.001$; **: $p < 0.01$; *: $p < 0.05$; N.S.: $p > 0.05$; a. the values with parenthesis are standard deviations.

[b]: The mean value in latter condition minus the mean value in former condition.

Wu et al.

Wu et al. Journal of Trust Management 2015 2:2, doi:10.1186/s40493-015-0013-5

For expertise, both the review count (F $(1,190) = 5.656$, $p < 0.05$) and review balance conditions (F(4, 190) = 35.906, $p < 0.001$) have significant main effects. Moreover, a significant interaction effect is observed (F $(4, 190) = 13.05$, $p < 0.001$). This result suggests that the advisor's expertise is perceived to be higher under low number of reviews condition than under high number of reviews condition (mean differences = 0.288, $p < 0.05$). And the values of perceived expertise are distributed differently across high and low number of reviews conditions. In low number of reviews condition, only the difference between means in conditions "R<<S" and "R<S" is negative (−0.36, but insignificant), suggesting that the perceived expertise linearly increases when review balance ranges from extremely positive to extremely negative. However, in the high number of reviews condition, the values of perceived expertise are distributed differently (an inverted-U shape). Especially when advisors have almost all negative reviews, they are perceived to be very low in expertise (see Table 5, repeated contrast of means between conditions "R<<S" and "R<S": −2.64, $p < 0.001$).

In line with study 1, study 2 supports H2 but rejects H1. The results from both study 1 and study 2 show that buyers might misattribute low trustworthiness to low expertise, and this case might happen when buyers check an advisor who has a high number of reviews.

## Study 3

The misattribution phenomenon found in study 1 and study 2 suggests that it is necessary to further explore the interplay between sub-dimensions of trust and expertise. Previous studies indicate that misattribution is usually a kind of affective response to a stimulus [56]. Similar to source credibility, trust is also a multifaceted variable, including both cognitive dimension and affective dimensions [22].

Cognitive trust is a kind of prediction based on people's accumulated knowledge gained through observation of trustee's behavior [22]. Affective trust is generated based on the positive emotions in the judgement process. Previous studies assume a positive impact of cognitive trust on affective trust because cognitive trust is a prerequisite for affective trust [57], [22]. Cognitive trust has clear distinctions with expertise [7]. However, affective trust may have close relationship with expertise because of buyer's misattribution. Therefore, we conducted study 3 to explore the relationships among affective trust, cognitive trust and expertise in high number of reviews condition.

## *Details of Experiment*

A survey-based experiment was conducted. Detailed content of measurable items are shown in Table 6. Three measureable items (AFF3, AFF4, AFF5) for trust are extracted from previous study [57], while others are self-developed. Self-developed measures were used because no relevant items can be found in previous studies, and these items were developed to fit our research context well. Items used to measure expertise are extracted from Ohanian [25].

**Table 6**: Results of measurement model in study 3

| Construct | Items | Content | C.R | C.A. | AVE | Loading | VIF |
|---|---|---|---|---|---|---|---|
| Cognitive trust | COG1 | I see no reason to doubt his motivation to write reviews | 0.94 | 0.91 | 0.75 | 0.77 | 2.26 |
| | COG2 | I think taking his review into consideration is a good decision | | | | 0.95 | 2.53 |
| | COG3 | I think I can rely on his reviews | | | | 0.77 | 2.79 |
| | COG4 | I think what he write in the reviews (pros and cons) is reasonable | | | | 0.92 | 2.79 |
| | COG5 | I think the review content and review activities make him a trustworthy advisor. | | | | 0.88 | 3.09 |

| Affective trust | AFF1 | I can feel his sincerity in writing reviews. | 0.93 | 0.92 | 0.76 | 0.84 | 2.77 |
|---|---|---|---|---|---|---|---|
| | AFF2 | I am confident that he writes reviews based on his real experience. | | | | 0.95 | 3.51 |
| | AFF3 | I feel comfortable about relying on him for my purchase decision. | | | | 0.87 | 1.79 |
| | AFF4 | I feel secure about relying on him for my purchase decision | | | | 0.83 | 3.38 |
| | AFF5 | I feel content about relying on him for my purchase decision | | | | 0.85 | 3.64 |
| Perceived expertise | EXP1 | Expert-not an expert | 0.93 | 0.91 | 0.74 | 0.88 | 2.94 |
| | EXP2 | Experienced-inexperienced | | | | 0.89 | 2.39 |
| | EXP3 | Knowledgeable-unknowledgeable | | | | 0.86 | 2.83 |
| | EXP4 | Qualified-unqualified | | | | 0.83 | 2.58 |
| | EXP5 | Skilled-unskilled | | | | 0.83 | 2.79 |

Note: S.D.: standard deviation. C.R.: Composite reliability. C.A.: Cronbach's alpha.

Wu et al.

Wu et al. Journal of Trust Management 2015 2:2, doi:10.1186/ s40493-015-0013-5

The experiment procedure is similar to the procedure in study 2. We invited 100 undergraduate students with Taobao purchase experience to take part in our experiment. The demographic information of participants is shown in Table 7. The number of participants meets the requirement of Partial Least Squares (PLS) analysis. Each participant was randomly assigned into one of the five review balance conditions with an advisors' review history containing 200 reviews. Participants were told to imagine that they were shopping in Taobao and need to judge the credibility of the advisor. Survey was provided as soon as the participants finished their judgement.

**Table 7**: Demographic information of participants

| Items | Mean | S.D. | Min | Max | Comment |
|---|---|---|---|---|---|
| 1. Age | 22.24 | 1.11 | 19 | 25 | |
| 2. Gender | 0.49 | 0.50 | 0 (female) | 1 (male) | Male:49; Female:51 |
| 3. How much Taobao purchase experience do you have? | 4.95 | 0.76 | 4 | 6 | 7 point scale (rarely-very frequently) |

Wu et al.

Wu et al. Journal of Trust Management 2015 2:2, doi: 10.1186/ s40493-015-0013-5

Structural equation modeling (SEM)-based PLS analysis was chosen to process survey data. This method was chosen in this study because it is suitable for exploratory study, and it requires neither large sample size nor multivariate normality of distribution [44]. We used WarpPLS 4.0 with bootstrapping to conduct PLS analysis. In line with other PLS softwares, the classic PLS algorithm was adopted.

## *Analysis and Result*

The analysis procedure is divided into two steps: test for measurement model and structural model. Table 6 shows the results of measurement model. All factor loadings were significant ($p < 0.001$), and ranged from 0.77 to 0.95. The composite reliability and Cronbach's alpha of each factor ranged from 0.91 to 0.94. All values of AVE are greater than 0.5. Finally, multi-collinearity is not a serious issue because the highest value of VIFs is only 3.64. These results indicate that our self-developed questions have good reliability and our survey data are suitable for further analysis.

In the test of structural model, first, age, gender and purchase experience are included as control variables. Results show that p values for these three variables are 0.06, 0.35 and 0.19. Therefore, no significant effects ($p > 0.05$) of control variables are found. As it is shown in Figure 3, the impact of cognitive trust on expertise is not significant (Beta $= 0.05$, $p > 0.05$). Cognitive trust has positive impact on affective trust (Beta $= 0.76$, $p < 0.001$), and affective trust positively influences expertise (Beta $= 0.45$, $p < 0.001$). The percentage of the variance explained ($R^2$) of affective trust and perceived expertise are 57% and 29%, indicating good explanation power.

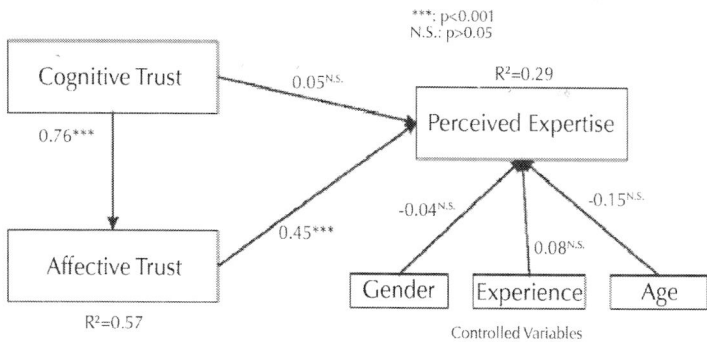

**Figure 3**: Results of structural model in study 3.

The results of study 3 confirm the assumption of misattribution from trustworthiness to expertise, and further suggest that affective trust plays a significant role in determining expertise.

# Summary and Discussions

In online marketplaces, an advisor's credibility is important because buyers rely on advisor's reviews to make purchase decision. An advisor's profile is a major way for buyers to assess advisor's credibility. A profile usually includes identity-related information and review history. Disclosure of identity-related information has been found to be helpful in supporting buyers' judgment, however, the impacts of the review history remains unclear. In this research, we investigated the effects of review balance, an important aspect of review history. Study 1 investigated how buyers perceive advisors' trustworthiness and expertise based on different review balances. The results support H2 and show that perceived trustworthiness distributes in an inverted U-shaped curve when review balance ranges from extremely negative to extremely positive. Advisors with almost all positive or negative reviews are perceived to be not trustworthy, while advisors who write mixed reviews are perceived to be trustworthy. This result is in line with psychological studies [45], suggesting that mixed positive and negative reviews could enhance buyers favorable judgement towards a target (a seller, a product or an advisor). The finding is also supported by data mining studies, which treat advisors with all negative reviews as unusual cases with low trustworthiness [30],[31],[34],[58],[59]. An unexpected result in study 1 is that perceived expertise does not decrease linearly when review balance ranges from extremely negative to extremely positive. Therefore H1 is rejected.

An advisor with almost all positive reviews might be seen as an easy-to-satisfy buyer. As the proportion of negative reviews increases, advisor's perceived strictness on evaluating product increases. However, an advisor with extremely high proportion of negative reviews is perceived to be low in expertise. This result implies that buyers might misattribute low trustworthiness to low expertise. Many trust-related misattributions have been mentioned in previous studies. For example, alcoholism, drug abuse, and mental illness among managers can harm employee's trust towards the organization [60]; people with positive emotions (e.g., happiness and gratitude) are more inclined to trust than people with negative emotions (e.g., anger, sadness) [20]. This phenomenon occurs because affective states, even if they are caused by unrelated events, usually serve as an information aid in people's judgement.

Study 2 addresses some limitations of study 1 by incorporating larger sample size and more variables. The results of study 2 are consistent with those found in study 1, and again reject H1 but support H2. Study 2 further suggests that buyer's misattribution behavior is more likely to happen under high processing effort condition (advisor with high number of reviews). Currently there is little evidence to support the direct relationship between stress and misattribution. However, processing a high number of reviews can cause low processing fluency, which then leads to negative affective states [61].

The result of study 3 shows that expertise is positively related to affective trust, while not significantly related to cognitive trust. It provides evidence to explain why low trustworthiness leads to low expertise. Previous studies, however, neglect the affective aspect of trust and argue that trustworthiness and expertise are clearly distinguishable [7].

# Implications

The results of this study yielded a couple of theoretical implications. First, previous studies on online marketplace mainly focus on the importance of advisors' review for potential buyers in evaluating the trustworthiness of sellers. They advocate that the existence of inconsistent reviews, rather than majority positive or majority negative reviews, better reflects the seller's credibility. However, they did not consider different credibility of advisors. This study explores how the advisor's profile signals credibility meaning to buyers. Second, a large amount of work on advisors' credibility focuses on static personal information (e.g., gender, hobbies). This study moves a step further to evaluate the impact of review balance shown in review history. It is worthwhile to explore review history since it can provide valuable information to judge advisors' credibility. Third, this study enriches extant knowledge about the relationship between trustworthiness and expertise. Previous studies mention that people can easily distinguish between trustworthiness and expertise [7], and in some experiment, manipulations of trustworthiness and expertise were not found to influence each other [62]. However, in this study, low expertise is found to easily be misattributed from low trustworthiness, especially when buyers face advisors with a high number of reviews.

This study also generates practical implications for the design of mechanisms to support credibility judgement. First, there are many ways to assigning trust value when buyers and sellers are strangers, including initializing trust values based on beta distribution, and incorporating social network attributions. We argue that assigning trust values should take subjective perception into consideration. The results of this study could serve as a reference for assigning credibility values of advisors. Second, some trust models compute advisors' trustworthiness based on the degree of consensus among advisors [36]. Such method might not be suitable in a marketplace with a high proportion of fake positive reviews, because these models assume other advisors are credible. The results of this study could be helpful to refine existing trust models by reducing the importance of consensus in considering trustworthiness. For example, a malicious advisor with all positive reviews might be judged in existing models as highly trustworthy because his reviews are in agreement with others, however, he will be considered to be less trustworthy in revised trust model.

## Limitations and Future Work

This study has five limitations, which affect the generalizability of our findings. First, although our research participants (mostly undergraduates) reflect a typical group of buyers in online marketplace, they cannot be representative of the whole consumer community. Moreover, our participants were required to have purchase experience and they were aware of unfair/review issues in online marketplace, therefore our findings cannot fully explain how new buyers perceive credibility of advisors with different review histories. We will extend our work by inviting participants with various backgrounds in future work. Second, in our experiment, an advisor's reviews for all sellers were listed together, but the differences (e.g., reputation) among different sellers were not considered. We argue that discarding sellers' difference does not significantly affect our result because it is unlikely for a buyer to further judge characteristics of sellers who are listed in an advisor's profile page. This issue will be considered in future work as a pretest before formal experiment. Third, our results cannot explain how buyers perceive an advisor who only has a few reviews. Buyers usually cannot make judgement based on a short review history list (e.g., only one or two reviews). Fourth, different online marketplaces have different characteristics. Our target platform (Taobao) has serious unfair rating/review problem, while this issue might not be a problem in other platforms. Therefore, buyers in Taobao are assumed to have more knowledge about identifying advisors with low credibility. Fifth, in real purchase, buyers usually have to judge a list of advisors, while our experiments (study 2 and 3) only required participants to judge one advisor. The judgement of a list of advisor might be affected by the sequence of the list (primacy effect: buyers can only remember the credibility of the first advisor) and the information overload (e.g., buyers only judge a few advisors in the list). In future study, we will aim at measuring trust attitude towards a seller by providing buyers with a list of advisors.

# AUTHORS' CONTRIBUTIONS

KW and ZN proposed initial idea of this paper. JV provided comments and helped KW and ZN with the formulation of research framework.

KW and ZN conducted study 1 and KW conducted study 2 and 3. KW and ZN drafted the paper. IA proofread the draft and provided comments and suggestions. KW revised this paper according to reviewers' comments. JV supervised the revising process, proofread revised paper and confirmed responses to review comments. KW was in charge of submitting the paper and corresponding with the editors of the journal and Springer Open Production Team. All authors read and approved the final manuscript.

# ACKNOWLEDGEMENTS

This work has been supported by NSERC through a Discovery Grant and a Discovery Accelerator Supplement Grant.

# REFERENCES

1.    Li M, Huang L, Tan C-H, Wei K-K: Helpfulness of online product reviews as seen by consumers: Source and content features. *Int J Electron Commer* 2013, 17(4):101-136.

2.    Lee J, Park D-H, Han I: The effect of negative online consumer reviews on product attitude: An information processing view. *Electron Commer Res Appl* 2008, 7(3):341-352.

3.    Park D-H, Lee J, Han I: The effect of on-line consumer reviews on consumer purchasing intention: The moderating role of involvement. *Int J Electron Commer* 2007, 11(4):125-148.

4.    Chua AYK, Banerjee S (2014) Understanding review helpfulness as a function of reviewer reputation, review rating, and review depth. Journal of the Association for Information Science and Technology

5.    Zhang Y, Bian J, Zhu W: Trust fraud: A crucial challenge for china's e-commerce market. *Electron Commer Res Appl* 2012, 12(5):299-308.

6.    Dellarocas C: Immunizing online reputation reporting systems against unfair ratings and discriminatory behavior. In *Proceedings of the 2nd ACM conference on Electronic commerce*. ACM, New York; 2000:150-157.

7.   Sprecker K: How involvement, citation style, and funding source affect the credibility of university scientists. *Sci Commun* 2002, 24(1):72-97.

8.   Chen Z, Yang J: Credit fraud control and credit system optimization on c2c marketplaces. In *Proceedings of the 42nd Hawaii International Conference on System Sciences*. Hawaii, HI, IEEE Computer Society Press, Washington; 2009.

9.   Lim Y-S, Van Der Heide B (2014) Evaluating the wisdom of strangers: The perceived credibility of online consumer reviews on yelp. Journal of Computer-Mediated Communication

10.  Park H, Xiang Z, Josiam B, Kim H: Personal profile information as cues of credibility in online travel reviews. *Anatolia* 2013, 25(1):13-23.

11.  Forman C, Ghose A, Wiesenfeld B: Examining the relationship between reviews and sales: The role of reviewer identity disclosure in electronic markets. *Inf Syst Res* 2008, 19(3):291-313.

12.  Ghose A, Ipeirotis PG: Estimating the helpfulness and economic impact of product reviews: Mining text and reviewer characteristics. *IEEE Trans Knowl Data Eng* 2011, 23(10):1498-1512.

13.  Heesacker M, Petty RE, Cacioppo JT: Field dependence and attitude change: Source credibility can alter persuasion by affecting message-relevant thinking. *J Pers* 1983, 51(4):653-666.

14.  Tintarev N, Masthoff J: A survey of explanations in recommender systems. In *IEEE 23rd International Conference on Data Engineering Workshop*. IEEE Computer Society Press, Washington; 2007:801-810.

15.  Rezaei S: Segmenting consumer decision-making styles (cdms) toward marketing practice: A partial least squares (pls) path modeling approach. *J Retail Consum Serv* 2015, 22:1-15.

16.  Mcallister DJ: The second face of trust:Reflections on the dark side of interpersonal trust in organizations. *Res Negotiation Organ* 1997, 6:87-111.

17.  Tarnanidis T, Owusu-Frimpong N, Nwankwo S, Omar M: Why we buy? Modeling consumer selection of referents. *J Retail Consum Serv* 2015, 22:24-36.

18. Park J, Gunn F, Han S-L: Multidimensional trust building in e-retailing: Cross-cultural differences in trust formation and implications for perceived risk. *J Retail Consum Serv* 2012, 19(3):304-312.

19. Willemsen LM, Neijens PC, Bronner F: The ironic effect of source identification on the perceived credibility of online product reviewers. *J Comput-Mediat Commun* 2012, 18(1):16-31.

20. Dunn JR, Schweitzer ME: Feeling and believing: the influence of emotion on trust. *J Pers Soc Psychol* 2005, 88(5):736.

21. Willemsen LM, Neijens PC, Bronner F, De Ridder JA: "Highly recommended!" the content characteristics and perceived usefulness of online consumer reviews. *J Comput-Mediat Commun* 2011, 17(1):19-38.

22. Johnson D, Grayson K: Cognitive and affective trust in service relationships. *J Bus Res* 2005, 58(4):500-507.

23. Heider F: *The psychology of interpersonal relations*. Psychology Press, New Jersey; 2013.

24. Pornpitakpan C: The persuasiveness of source credibility: a critical review of five decades' evidence. *J Appl Soc Psychol* 2004, 34(2):243-281.

25. Ohanian R: Construction and validation of a scale to measure celebrity endorsers' perceived expertise, trustworthiness, and attractiveness. *J Advert* 1990, 19(3):39-52.

26. Sparks BA, Browning V: The impact of online reviews on hotel booking intentions and perception of trust. *Tour Manag* 2011, 32(6):1310-1323.

27. Huang L, Tan C-H, Ke W, Wei K-K (2014) Do we order product review information display? How? Information & Management

28. Khare A, Labrecque LI, Asare AK: The assimilative and contrastive effects of word-of-mouth volume: An experimental examination of online consumer ratings. *J Retail* 2011, 87(1):111-126.

29. Mudambi SM, Schuff D: What makes a helpful online review? A study of customer reviews on amazon.com. *Manag Inf Syst Q* 2010, 34(1):185-200.

30. Metzger MJ, Flanagin AJ, Medders RB: Social and heuristic approaches to credibility evaluation online. *J Commun* 2010, 60(3):413-439.

31.    Chou SY, Picazo-Vela S, Pearson JM: The effect of online review configurations, prices, and personality on online purchase decisions: a study of online review profiles on ebay. *J Internet Commer* 2013, 12(2):131-153.

32.    Amabile TM: Brilliant but cruel: perceptions of negative evaluators. *J Exp Soc Psychol* 1983, 19(2):146-156.

33.    Ashforth BE, Mael F: Social identity theory and the organization. *Acad Manag Rev* 1989, 14(1):20-39.

34.    Jindal N, Liu B, Lim E-P: Finding unusual review patterns using unexpected rules. In*Proceedings of the 19th ACM international conference on Information and knowledge management*. ACM, New York; 2010:1549-1552.

35.    Ziegler C-N, Golbeck J: Investigating interactions of trust and interest similarity. *Decis Support Syst* 2007, 43(2):460-475.

36.    Zhang J, Cohen R: Evaluating the trustworthiness of advice about seller agents in e-marketplaces: a personalized approach. *Electron Commer Res Appl* 2008, 7(3):330-340.

37.    Zhu Y, Zhang W, Yu C: Detection of feedback reputation fraud in taobao using social network theory. In *2011 International Joint Conference on Service Sciences (IJCSS)*. IEEE Computer Society Press, Washington; 2011:188-192.

38.    Popper B (2014) Alibaba has more sales than amazon and ebay combined, but will americans trust it? http://www.theverge.com/2014/5/7/5690596/meet-alibaba-the-ecommerce-giant-with-more-sales-than-amazon-and-ebay.

39.    Clemes MD, Gan C, Zhang J: An empirical analysis of online shopping adoption in beijing, china. *J Retail Consum Serv* 2014, 21(3):364-375.

40.    Moneyweek (2011) Gray industry cluster parasitism: Four complicated interest chains on taobao. http://finance.stockstar.com/MS2011050300000899.shtml.

41.    Cnnic (2014) 2013 statistical report on chinese internet shopping https://www.cnnic.net.cn/hlwfzyj/hlwxzbg/dzswbg/201404/t20140421_46598.htm.

42.    Sen S, Lerman D: Why are you telling me this? An examination into negative consumer reviews on the web. *J Interact Mark* 2007, 21(4):76-94.

43. Pavlou PA, Dimoka A: The nature and role of feedback text comments in online marketplaces: Implications for trust building, price premiums, and seller differentiation. *Inf Syst Res* 2006, 17(4):392-414.

44. Kallweit K, Spreer P, Toporowski W: Why do customers use self-service information technologies in retail? The mediating effect of perceived service quality. *J Retail Consum Serv* 2014, 21(3):268-276.

45. Ein-Gar D, Shiv B, Tormala ZL: When blemishing leads to blossoming: the positive effect of negative information. *J Consum Res* 2012, 38(5):846-859.

46. Berger J, Sorensen AT, Rasmussen SJ: Positive effects of negative publicity: when negative reviews increase sales. *Mark Sci* 2010, 29(5):815-827.

47. Xie H, Miao L, Kuo P-J, Lee B-Y: Consumers' responses to ambivalent online hotel reviews: the role of perceived source credibility and pre-decisional disposition. *Int J Hosp Manag* 2011, 30(1):178-183.

48. Nevo D, Chan YE: A delphi study of knowledge management systems: scope and requirements. *Inform Manag* 2007, 44(6):583-597.

49. Senecal S, Nantel J: The influence of online product recommendations on consumers' online choices. *J Retail* 2004, 80(2):159-169.

50. Cohen J (2013) Statistical power analysis for the behavioral sciences. Academic, Routledge

51. Liqiong D, Poole MS: Affect in web interfaces: a study of the impacts of web page visual complexity and order. *MIS Q* 2010, 34(4):711-A710.

52. Kim DJ, Ferrin DL, Rao HR: A trust-based consumer decision-making model in electronic commerce: The role of trust, perceived risk, and their antecedents. *Decis Support Syst* 2008, 44(2):544-564.

53. Mackenzie SB, Podsakoff PM, Podsakoff NP: Construct measurement and validation procedures in mis and behavioral research: Integrating new and existing techniques. *MIS Q* 2011, 35(2):293-A295.

54.  Mason CH, Perreault WD Jr: Collinearity, power, and interpretation of multiple regression analysis. *J Mark Res* 1991, 28(3):268-280.

55.  Cenfetelli RT, Bassellier G: Interpretation of formative measurement in information systems research. *MIS Q* 2009, 33(4):7.

56.  Payne BK, Hall DL, Cameron CD, Bishara AJ (2010) A process model of affect misattribution. Bulletin, Personality and Social Psychology

57.  Sun H: Sellers' trust and continued use of online marketplaces. *J Assoc Inf Syst* 2010, 11(4):2.

58.  Mukherjee A, Liu B, Wang J, Glance N, Jindal N: Detecting group review spam. In*Proceedings of the 20th international conference companion on World Wide Web*. ACM, New York; 2011:93-94.

59.  Lim E-P, Nguyen V-A, Jindal N, Liu B, Lauw HW: Detecting product review spammers using rating behaviors. In *Proceedings of the 19th ACM international conference on Information and knowledge management*. ACM, New York; 2010:939-948.

60.  Speller JL: *Executives in crisis: Recognizing and managing the alcoholic, drug-addicted, or mentally ill executive*. Jossey-Bass, San Francisco; 1989.

61.  Mosteller J, Donthu N, Eroglu S: The fluent online shopping experience. *J Bus Res* 2014, 67(11):2486-2493.

62.  Wiener JL, Mowen JC: Source credibility: on the independent effects of trust and expertise. *Adv Consum Res* 1986, 13(1):306-310.

# Chapter 5

# Enhancing Science and Technology Cooperation between the EU and Eastern Europe as Well as Central Asia: A Critical Reflection on the White Paper from a S&T Policy Perspective

Klaus Schuch[1], George Bonas[2], and Jörn Sonnenburg[3]

[1]Centre for Social Innovation, Linke Weinzeile 246, Wien, 1150, Austria

[2]International Centre for Black Sea Studies and National Hellenic Research Foundation, 4 Xenophontos Street, Athens, 10557, Greece

[3]International Bureau of the Federal Ministry of Education and Research, Heinrich-Konen-Street 1, Bonn, 53227, Germany

# ABSTRACT

This article reflects the main findings of the 'White Paper on opportunities and challenges in view of enhancing the EU cooperation with Eastern Europe, Central Asia and South Caucasus in Science, Research and Innovation', which was released in April 2012, from a science and technology (S&T) internationalisation policy perspective. In the 'Internationalisation of R&D from an S&T policy perspective' section of this article, the ongoing discourse on internationalisation of research and development (R&D) is discussed from an S&T policy perspective. In the 'S&T cooperation between the EU and Eastern Europe as well as Central Asia since the early 1990s' section, the development of S&T cooperation between the EU and EECA is described as a historical snapshot since the early 1990s. In the 'Recent S&T internationalisation efforts of Eastern European and Central Asian countries' section, special emphasis is given to the current EECA countries' dispositions towards R&D internationalisation. For a structured overview, the EECA region is disaggregated in three subregions, namely, (a) Russian Federation, (b) Eastern European countries (without Russia) and (c) Central Asian countries. To better position the R&D internationalisation policies of the region under scrutiny within the overall state-of-the-art of S&T, the 'Current state of S&T in the Eastern European and Central Asian countries' section compares main S&T indicators of the EECA countries. The 'The White Paper recommendations in the light of international S&T cooperation policy objectives' section finally condenses the major recommendations elaborated during the White Paper consultation process and puts them into the context of international S&T cooperation policy. The question is raised on what international cooperation can contribute to improving S&T in the EECA region and which approaches are deemed most adequate to support this. The analysis shows that most recommendations suggested in the White Paper directly target the S&T policy (delivery) system, which is put into an explicit actor's role. Science diplomacy is the identified predominant driver for deepening international R&D cooperation with the EECA region. The main instruments used are international dialogue, exchange and learning platforms, which are supported by the European Commission according to the EU's subsidiarity principle. Other S&T internationalisation policy objectives play a role too, especially if a more regionally differentiated perspective is taken into account.

# BACKGROUND

## Internationalisation of R&D from an S&T Policy Perspective

Internationalisation of research and development (R&D) is a phenomenon which received attention only recently and for which insufficient data and data comparability are characteristic elements (Edler and Flanagan [2009]; OECD [2005]; Schuch [2011]). The academic focus on R&D internationalisation is predominantly actor-focussed, with a strong emphasis on private, industry-driven R&D (Cantwell [1995]; Dalton and Serapio [1999]; Narula and Zanfei [2004]; OECD [2008]; Verbeek and Shapira [2009]) and with secondary emphasis on public R&D organisations. The fundamental typology of Archibugi and Michie ([1995]), who differentiate between exploitation, cooperation and generation within international R&D cooperation, is an influential example for this actor-centred approach, which puts industry-driven R&D, especially of multinational enterprises, in the centre of investigations. The role of S&T policy for R&D internationalisation is regarded primarily as an accompanying 'enabling' or - at least - 'preventing' framework. The enabling function comprises the development of stimulating incentives or support programmes such as cross-border R&D programmes and/or the openness of national programmes and projects (Edler et al. [2002]), while the preventing function primarily concerns the protection of intellectual property at an international scale. Above all, however, the main task of public S&T policy towards internationalisation of R&D is to keep its own house clean, i.e. to be an attractive place for conducting R&D and, thus, for attracting R&D inflows from abroad too (Verbeek and Shapira [2009]).

In 2008, a Comité de la Recherche Scientifique et Technique (CREST) [a] working group made a comprehensive attempt to analyse public S&T policies of 21 European countries [b] towards R&D internationalisation by placing R&D and innovation policy in an actor's role (Sonnenburg et al. [2008]). This study clearly revealed that in most countries, which participated in this working group, national S&T policies actively started to deal with internationalisation of R&D, not just to let it happen, but to support it and even to direct it. Examples for this proactive understanding are incentives to attract inward corporate and institutional R&D, to participate in cross-border research programmes, to invest in joint R&D labs abroad and to support the mobility of researchers or the coordination of R&D internationalisation policies among European Union (EU) Member States and countries associated to the EU RTD Framework Programme towards third countries. Basically, two different sets of R&D internationalisation objectives could be distinguished: an intrinsic dimension, which puts goals into the centre of public S&T policy that directly aim to substantiate S&T (e.g. trough enabling R&D cooperation among the best researchers globally or to find joint solutions for large-scale R&D infrastructures which cannot be financed by a country on its own); and an extrinsic dimension, which rather focuses on goals that are meant to support other policies (e.g. facilitation of access to foreign markets through standard settings or research for development to assist technical development cooperation). The CREST study, however, also revealed that interventionist approaches of (primarily national) S&T policy towards R&D internationalisation addressed first-of-all public R&D organisations and agencies. In addition, some measures of more generic nature were triggered by public S&T policy while an explicit focus on the private business-enterprise sector was rather limited but progressively taken up in the academic discourse (Edler[2008]; Rama [2008]). Little is yet known about the drivers of public R&D organisations to participate in international R&D cooperation, but Edler ([2007]) identified in German public research organisations the following main motivations to internationalise, which might also be of relevance in other European countries: firstly, access to and utilisation of excellent and complementary knowledge abroad and, secondly, to secure funding (mainly via EU sources), followed by building up of reputation and visibility of the public R&D organisation. In contrast to the industrial sector (Sachwald [2008]), cost advantages did hardly account.

[a]CREST (since 26 May 2010 renamed into ERAC: European Research Area Committee) is a strategic policy advisory body whose function is to assist the European Commission and the Council of the European Union in performing the tasks incumbent on these institutions in the sphere of research and technological development.

[b]Austria, Belgium, Czech Republic, Denmark, Germany, Greece, Finland, France, Ireland, Island, The Netherlands, Norway, Poland,

Portugal, Romania, Slovenia, Spain, Sweden, Switzerland, Turkey and United Kingdom.

These motives overlap with the main drivers of R&D internationalisation from a public policy perspective, which were identified by the CREST working group and which were confirmed by Boekholt et al. ([2009]), who included in their comparative study also policy examples from non-EU countries. Basically, the CREST working group identified the following objectives that drive R&D internationalisation from an S&T policy perspective:

- Quality acceleration and excellence
- Market and competition
- Resource acquisition
- Cost optimisation
- Global or regional development
- Science diplomacy

Different rationales are guiding these objectives: the rationale behind the *quality acceleration and excellence objective* is primarily an intrinsic one that assumes that international R&D cooperation improves the domestic science base, leading to faster and improved scientific progress as well as enhanced scientific productivity, and is also supportive for the professional advancement of the involved researchers (e.g. trough joint publications in acknowledged international journals). Behind this assumption stands the idea that only the 'best' (institutions and/or researchers) succeed also in international competitive procedures.[c] The rationale behind the extrinsic *market and competition objective* is to support the market entry of domestically produced technologies/innovations abroad as well as to support the access to and a quick uptake of technologies produced abroad within the domestic economy. Here, absorption capacities and the availability of efficient spill-over mechanisms are of importance. The rationale behind the *resource acquisition objective* overlaps partly with the two major objectives mentioned before. The access to information, knowledge, technology and expertise as well as to singular equipment/ facilities and materials are in the focus. However, resource acquisition is not limited to different codified and tacit dimensions of technology transfer but extends to brain gain, gaining of solvent students (for universities) and increasingly also gaining research funds from abroad or from multilateral or international sources. The *cost optimisation objective* from a public S&T policy focus does not primarily mean to use cost arbitrages (e.g. lower wages in a foreign country) as this might be a rational argument of the corporate sector, but rather focuses on cost sharing approaches to create critical mass in a certain science arena, e.g. to establish large-scale research infrastructures, and it also includes the rational of risk sharing. The assumption behind the *global or regional development objective* is the comprehension that many risks have no frontiers (e.g. infectious diseases or climate change) or cannot be solved without international cooperation and solidarity (e.g. Millennium Development Goals) and, thus, have to be tackled through international R&D collaboration (e.g. research for development). Also, the *science diplomacy objective* often refers to global challenges and to development cooperation agendas. Fundamentally, it has two main rationales: firstly, to support through R&D cooperation other external policy dimensions in terms of science for diplomacy (e.g. non-proliferation of mass destruction weapons through keeping former weapon researchers busy with civilian R&D projects) and, secondly, to promote its own science base abroad in support of other objectives already mentioned above (e.g. to attract 'brains' or to promote a general quality trademark like 'made in Germany').

cThis assumption can, however, be challenged. A deliberation on this is provided by Schuch (2011).

Public S&T policies towards R&D internationalisation have both a strong 'inward' dimension, which is to reinforce the domestic science base through attraction of foreign resources (e.g. human resources, knowledge or foreign funds), and a strong 'outward' dimension in linking domestic actors to knowledge produced abroad (Edler and Boeckholt [2001]). Another channel for absorption is to integrate foreign actors into cooperation programmes (Verbeek and Shapira[2009]). The latter aspect of R&D internationalisation has been taken up by the European Commission in the European Framework Programmes (FPs) for Research, Technological Development and Demonstration (RTD). The most recent communication of the European Commission (EC) on internationalisation, which gives orientation for FP7, puts the issue of excellence through competition (or better: co-opetition[d]) in the forefront: 'Excellence in research stems from competition between researchers and from getting the best to compete and co-operate with each other. A crucial way to achieve this is […] to work together across borders' (European Commission [2008], p. 4). This stems from the belief that the EU does not claim to be a self-sufficient entity in the realm of S&T and innovation, but that both Europe's knowledge resource (e.g. human capital) and its role in the global economy will be increasingly shaped by its ability to source knowledge internationally and to adapt it for its own use. In the EC's green paper on the European Research Area (ERA), in which six key features were outlined to structure the ERA, the last key feature addresses the opening of the ERA to the world with special emphasis on neighbouring countries and a strong commitment in addressing global challenges with Europe's partners (European Commission [2007]).

ᵈdefined as the duality of competition and cooperation expressed on competitive markets.

While a development policy approach ('research for development') was the main driver in FP3 for the opening up of the European Framework Programme in dedicated thematic niches (e.g. agro-food related R&D), this approach was soon complemented by a more explicit S&T diplomacy approach towards Eastern Europe and the New Independent States of the Former Soviet Union (NIS) after the breakdown of the communist hegemony and the collapse of the Soviet Union. On one hand, this S&T diplomacy approach was driven by the geopolitical concern to bring Central European Candidate Countries closer to the EU and it's upcoming research area (and then to integrate them). On the other hand, this approach was driven by a neighbourhood-oriented stabilisation policy with a special focus on non-proliferation of mass destruction weapons in the NIS. In the following chapter, the development of S&T cooperation relations between the EU and the Eastern European and Central Asian (EECA) countries is summarised in a historical perspective.

# RESULTS AND DISCUSSION

## S&T Cooperation between the EU and Eastern Europe as Well as Central Asia since the Early 1990s

International science and technology cooperation between the EU, its member states and the post-socialist Eastern European and Central Asian countries began soon after the collapse of the hegemonic communist system in early 1990s. At community level, the European Commission pursued and supported collaborative trans-European R&D efforts whose main aims were to safeguard and strengthen the S&T potential in the EECA partner countries by orienting research towards new socioeconomic needs of the transition countries, to prevent proliferation of military-relevant knowledge and to generate and disseminate new scientific and technological knowledge by encouraging enterprises and research institutes from the East and West to carry out joint research projects and to organise technology transfer under the European FPs for RTD. In FP4, which lasted from 1994 to 1998, the PECO and COPERNICUS schemes were the EC's main mechanism for stimulating S&T cooperation between researchers from the EU and researchers from the Central European Candidate Countries as well as from the New Independent States of the Former Soviet Union (NIS). The EC spent 241 million ECU in FP4 under those schemes for more than 500 projects involving 3,286 participants (i.e. research entities from the university sector, industrial sector or non-university research sector) (European Commission [1999]). Forty-six percent of the approved projects involved only participants from the Central European Candidate Countries and the EU15 (as well as countries associated to the FP at that time). Twenty-one percent involved only participants from the NIS and the EU15 (plus associated states), and 33 % involved participants from all three major regions. Among all target countries, Russia had the most participants under COPERNICUS (FP4) followed by five Central European Candidate Countries, while Ukraine ranked 7th and Belarus, 10th (Schuch [2005]).

In addition to COPERNICUS, scientific co-operation with the NIS was also supported by the International Association for Cooperation with Scientists from the former Soviet Union (INTAS), which was established as an international association under Belgian law in 1993 (INTAS [1995]) by the European Commission, the EU member states and countries associated to the EU RTD Framework Programme. Until the turn of the millennium, more than 20,000 individual scientists from the NIS had been involved in approximately 2,000 INTAS projects. From 1993 to 1998, the association's budget totalled 121 million to which another overall budget of 75 million (European Commission [2000]) has been added from 1998 until the end of 2002. While COPERNICUS was governed by the EC directly, the INTAS members had a strong influence on the governance of INTAS.

To complete the picture, it has to be noted that the EU together with the USA, Japan and the Russian Federation established the International Science and Technology Centre (ISTC) in Moscow in 1992, whose membership enlarged since then. The primary aim of ISTC is to offer opportunities to scientists working in the former Soviet Union's military research programmes to redirect their skills towards civilian research and to prevent the expertise and technologies of weapons of mass destruction from proliferating. In 1993, a similar agreement was signed between the USA, Canada, Sweden and Ukraine to establish the Science and Technology Centre of the Ukraine in Kiev. Subsequently, Sweden was replaced by the European Union.

Using these different co-operation instruments focusing primarily on joint R&D projects, loose individual contacts among researchers from EU and EECA countries were strengthened and transformed into more sustainable scientific relations. Although the disbursed foreign grants could not counterbalance the dramatic drop in R&D expenditures in the NIS, they at least had a positive impact on the diffusion of new approaches and methodologies as well as standards and model practices (Le Gohebel et al. [2011]).

This first phase of the EU's R&D cooperation with the EECA region coincided with heavy sociocultural transformation processes, which forced the research systems in Eastern Europe and Central Asia to change, not due to emerging new and proactive stimuli or ideas about efficient national innovation systems, but mainly as a result of the severe socioeconomic transition crisis in which S&T - despite some lip service - was usually not treated as a priority policy area in the countries concerned (Schuch [2005]). Although the decline of the educational and research systems in the region under scrutiny was already apparent during the last years of the old regime, the downturn in economic activity during the first phase of the transformation process and the beginning of restructuring was accompanied by an accelerated winding down of the research capacities. It is worthwhile noting, however, that in times of state-socialism, statistics tended also to overestimate R&D in comparison with OECD calculations (Godin [2001]; Gokhberg et al. [1999]).

Evidently, until the turn of the millennium, R&D cooperation between the EU and the Eastern European and Central Asian countries was predominately driven by science diplomacy rationales (stabilisation of S&T systems of neighbouring countries; non-proliferation of military know-how; support to foreign policy). Although some centres of excellence were recognised (especially in the fields of space research, materials and nuclear research), only Russia and Ukraine were, from the very beginning, considered as level-playing field partners, with whom international S&T cooperation has had a prevailing intrinsic research value in terms of achieving excellence through R&D cooperation.

The drivers of, and as a consequence, the instruments for international S&T cooperation between the EU and EECA countries started to gradually change at the beginning of the new millennium. This has been caused by several factors, such as the following:

- The economic recovery of most EECA countries
- A reclaimed self-confidence of Russia
- Consolidated structural changes in the national innovation systems in some EECA countries
- The emergence of new global players within the S&T arena (e.g. BRICS)
- The 'eastern' enlargement of the EU
- A general opening of FP6 and FP7 towards international partner countries
- An enhanced deliberation and importance gain of international S&T cooperation policy in general, leading - among other things - to more differentiated and targeted international S&T cooperation approaches

INTAS became one of the fist 'victims' of this reorientation. Although the programme proved to be successful (Idenburg et al. [2004]), it could not be adapted to a changing policy environment and ceased its operations at the end of the first decade of the new millennium. This was especially disadvantageous for the economically less advanced EECA countries since they hardly had capacities to withstand the higher competition exercised within the sixth and seventh FPs, which generally opened up for participation of international partner countries. Although the interest of Central Asian institutions in participating in FP7 is broad, in fact, only 34 institutes succeeded in different S&T projects within FP7 (data until May 2011). These participations were supported by the EC with a mere budget of 1.7 million. With 17 participations, Kazakhstan was the strongest partner from this region. Discussions with policy makers from Central Asia revealed that the European Framework Programme for RTD is rather a distant concept to them, while INTAS is still more present in their heads. A reason for this lack of awareness is also the underdeveloped FP7 National Contact Point (NCP) system in Central Asia. With 14 thematic NCPs and 1 national coordination office, Kazakhstan has the most developed NCP system in the region. A similarly advanced structure can only be found in Uzbekistan (13 NCPs). Overall, however, the existing FP7 national information points in the region are not directly supported financially by their national governments (Sonnenburg et al. [2012]).

The participation of research teams from Russia in successful projects supported under the European Framework Programmes for RTD differs significantly from all other EECA countries. Until the beginning of FP7, Russia has had consistently the highest project participation among the group of all 'third countries'. Its leading status as a preferential third partner country within FP7 is only contested by the USA. Under the framework of FP7, Russia, which concluded a first S&T agreement with the European Commission already in 1999, implements several 'coordinated calls' with the EU, which are jointly defined and funded. Since 2001, S&T agreements between the EU and Russia are also in place for EURATOM covering fission as well as fusion-oriented research.

The participation of the other Eastern European countries in FP7 lies - generally speaking - in between the one of Russia and that of the Central Asian countries. Up until the end of 2010, the majority of countries had a quite limited number of successful proposals, and the EC funding for the Eastern European countries' participants (except Russia) under FP7 ranges between 1 to 3 million per country. The only exception is Ukraine which had 103 successful proposals with an EC contribution reaching approximately 12 million. All Eastern European countries have a developed NCP structure in place to support domestic researchers in their aspirations to succeed in the competitive FP7 calls for proposals. In some countries, the NCPs are financially supported by the national authorities (e.g. Moldova, Ukraine). In some other countries, NCPs are not directly funded (Armenia, Belarus, Georgia).

# Recent S&T Internationalisation Efforts of Eastern European and Central Asian Countries

International S&T cooperation with the EECA countries, however, does not only occur under the European Framework Programmes for RTD. A variety of different policies and instrument are in place to substantiate international R&D cooperation. In order to provide a structured overview, we disaggregate the EECA region in three distinct subregions:

1.   Russian Federation
2.   Eastern European countries (without Russia) and
3.   Central Asian countries

Information provided in this section, if not indicated differently, is taken primarily from the White Paper (Sonnenburg et al. [2012]) and the analytical deliverables produced under the INCO-NET - EECA project, which can be accessed through the internet.[e]

[e]http:// http://www.inco-eeca.net/en/119.php.

# International S&T Cooperation in the Russian Federation

Enhancing internationalisation of the R&D sector has been identified by Russian policy makers as one important aspect for improving the quality and results of the Russian R&D system. Internationalisation in Russia, however, starts from a low level. Still, many R&D organisations are isolated from each other and from the outside world. Data on co-publication show that the USA and some EU countries (Germany, France, UK and Italy) are the top collaborating partners. Co-operation with China and South Korea is quickly increasing. A few Russian R&D programmes are also open for participation of EU researchers.[f] Main access obstacles are a lack of information about Russian research programmes, linguistic barriers and financial and legal issues.

[f]See http://www.access4.eu/index.php for more information

Russia has bilateral agreements and programmes in place with many states all over the globe. Since 1991, the USA have always been an important partner and among the first and largest investors in the Russian science and technology. The EU is another important partner for Russia's R&D internationalisation efforts. Russia has concluded bilateral S&T agreements with a broad range of EU member states and countries associated to the FP. At the level of research organisations, especially the Russian Academy of Sciences has a well-stocked network of cooperation agreements. Agreements have also been established at the level of research funds.

Findings of a survey conducted under the ERA.NET RUS project proved that bilateral international cooperation focuses on basic research. The most frequently used instrument is mobility support. Thus, not surprisingly, the budgets linked to bilateral agreements are mostly limited with annual amounts usually below 1 million. Most recent trends, however, show a shift from mobility towards more substantial R&D projects, a higher share for supporting applied research and innovation and an evolution from bilateral towards multilateral schemes (Kougiou et al. [2010]).

Russia is still intensively connected to its neighbouring countries in EECA at different cooperation levels. At the multinational level, the most important is the recently adopted 'Intergovernmental Programme for Innovation Cooperation of the Commonwealth of Independent States'.[g] Bilateral S&T agreements have been concluded with all EECA countries except Turkmenistan.[h] R&D cooperation within the Commonwealth of Independent States (CIS) is facilitated by the fact that Russian is considered as *lingua franca* among the scientific communities. In addition to the strong traditions and ties within the CIS, R&D cooperation with other Asian countries rapidly increases. The Russian Fund for Basic Research for instance regularly runs joint calls with the Japanese Society for the Promotion of Science, the State Fund for Natural Sciences of China and with the Indian Department of Science (Spiesberger [2008]).

[g]http:// http://rs.gov.ru/topic/185

[h]Taken from http://mon.gov.ru/work/mez/dok/1075/

Russian scientists participate also in projects launched under the European initiatives COST and EUREKA. Among all non-COST member countries, Russia has the highest participation in COST actions. Russian participation in EUREKA, however, is comparatively low, which confirms the limited innovation capacities of the country.

The latest joint EU-Russia initiative is a 'partnership for modernization', agreed in spring 2010. It includes cooperation in R&D and innovation. Regarding the latter, certain emphasis is on aligning technical regulations and standards between the EU and Russia and on enforcing IPR.

# International S&T Cooperation in Eastern European Countries (Without Russia)

The official S&T policy of all Eastern European neighbourhood countries acknowledges the importance of strengthening international R&D cooperation. Provisions (articles, paragraphs etc.) are included in the respective national legislations (e.g. Armenia: *Law on Scientific and Technological Activity, the Strategy on Development of Science* and Action Plan 2011–2015; Georgia: *Law on Science and Technologies and their Development*; Moldova: *Code On Science and Innovation*; *Moldova Knowledge Excellence Initiative* Action Plan 2008; Ukraine: *National Indicative Programme 2011–2013*). International S&T cooperation, for example, has got a special line in the Belarusian R&D state budget reserving 3% to 4% for international R&D activities annually. However, there is no distinct policy document referring to the issue of international R&D cooperation in any country.

Some of the national R&D programmes are open to foreign researchers, but usually, funds are provided only to domestic researchers.

The Eastern European countries signed bilateral agreements mainly with other CIS countries and countries of the EU. Some countries have also signed agreements with non-EU countries such as USA (Armenia), Argentina (Armenia), China (Armenia, Belarus, Moldova), India (Armenia, Belarus) and Venezuela (Belarus). Moreover, bilateral agreements have also been signed by research institutions (mainly the National Academies of Sciences) with similar counterparts abroad.

In addition to the national programmes, there are also a number of bilateral programmes in force involving other national bodies as well research organisations and centres. Examples are the following:

*Collaborative Programme* between CNRS, France and the State Committee of Science of the Republic of Armenia

The *Science and Technology Entrepreneurship Programme* between CRDF, USA and Georgian organisations

The collaborative calls between the Academy of Sciences of Moldova (ASM) and the Russian Foundation for the Humanities, as well as between the ASM and the German Federal Ministry of Education and Research

Several joint programmes of Belarus and Russia, e.g. the family of programmes for developing supercomputers – 'SKIF' (2000 to 2004), 'TRIADA' (2005 to 2008) and 'SKIF-GRID' (2007 to 2010) - with its follow-up, 'ORBISS' (2012 to 2015)

The regional cooperation within the EECA region is based on numerous bilateral inter-governmental agreements as well as on agreements between specific research institutions (academies, universities, research centres). The collaboration with Russia is characterized by the highest indices (e.g. in Belarus, 55% of the Academy's international projects are carried out with Russia). Overall, regional cooperation is mainly driven by personal or institutional links often inherited from Soviet times. In addition, regional cooperation also benefits from cross border programmes under the European Neighbourhood Policy Instrument (ENPI) (especially the *Black Sea Cross Border Cooperation Programme 2007–2013* and the *Black Sea Basin Joint Operational Programme 2007–2013*). Also, important for fostering the regional cooperation in S&T is the participation of almost all Eastern European countries in regional organisations such as the *Black Sea Economic Cooperation* and/or the *Organization for Democracy and Economic Development*(GUAM) which provide forums for political dialogue in various sectors including S&T. Within ENPI, however, S&T is not seen as a priority area for funding as such but can be supported only for regulatory reform and capacity-building activities (as it is the case with the operation of the Joint Support Office of the EC Nuclear Safety Programme for Ukraine).

According to the EU's *Competitiveness and Innovation Framework Programme* (CIP) regulations, this programme is open to third countries too. From the Eastern European neighbourhood countries, Armenia, Moldova and Ukraine participate in the *Enterprise Europe Network* of CIP[i] (a network of regional consortia providing integrated business and innovation support services for small and medium-sized enterprises (SMEs)) without receiving financial support from the programme. In addition, Moldova and Ukraine participate in the *Intelligent Energy Agencies initiative* of CIP again without financial support from the programme. All other Eastern European neighbourhood countries have not been involved yet in CIP.

[i]EEN Members: http://www.enterprise-europe-network.ec.europa.eu/about/branches

# International S&T Cooperation in the Central Asian Countries

International cooperation plays an increasingly acknowledged role in the implementation of the national S&T strategies in all Central Asian countries. International relations are usually regulated through presidential decrees (Uzbekistan, Tajikistan) or through the current laws on science (Kirgizstan, Kazakhstan, Tajikistan, and the law 'On the Status of Scientists' in Turkmenistan). The main national objectives of the Central Asian countries regarding international S&T cooperation include the following aspects: (a) exchange of S&T knowledge, (b) financial and technical support and (c) creation of joint research centres and organisations. The Kazakh State Programme 'The Path to Europe 2009-2011' is the only explicit international strategy established at national level. The aim of this programme is to bring the Republic of Kazakhstan to an advanced level of strategic partnership with leading European countries, especially in technological important fields like energy and transport, cooperation with SMEs as well as in social sciences and humanities.

The number of national programmes in Central Asia open to foreign researchers is significantly low. In Kazakhstan, the new 'Law on Science' encourages the participation of foreign researchers in national calls for proposals. Turkmenistan allows foreign participation in national programmes as part of technical assistance, but currently, the foreign participation in scientific activities includes mainly scientists from EECA, but not from other European countries.

Each Central Asian country has signed a few bilateral S&T agreements with different EU member states, for example, the *Agreement between the Government of Kazakhstan and the Government of Italy on Cultural and Scientific Cooperation* (11 May 2000), or the agreements between the government of Kazakhstan and the governments of Latvia and Estonia on economic and scientific-technical cooperation (March 2006). Aside from the EU, the countries have built up formal scientific relations with China (Kyrgyzstan), South Korea (Uzbekistan), USA (Uzbekistan, Tajikistan) or Afghanistan, Iran, Pakistan and India (Tajikistan). They traditionally cooperate very closely with the other countries of the former Soviet Union like Armenia, Belarus, Russia, etc. Besides government level agreements, bilateral collaboration is established also at the level of research-performing organisations, such as the national academies of science, state research centres and universities.

A considerable number of S&T cooperation agreements have been signed with the neighbouring countries in the years immediately after independence. Russia still remains the main S&T partner of the Central Asian countries. However, among others, the political situation in the region (e.g. conflicts in Tajikistan, Kyrgyzstan) influences strongly the scientific cooperation. Overall, the regional cooperation is still driven by the past (meaning Soviet) personal or institutional links, although also new initiatives emerged in the last few years. A good example for an existing regional approach is the *University of Central Asia* which operates in three countries of the region, i.e., Kazakhstan, Kyrgyzstan and Tajikistan. Further examples are as follows:

The *Eco-Regional Programme for Sustainable Agricultural Development* in Central Asia and the Caucasus,[j] a consortium of eight national agricultural research centres, eight centres of the*Consultative Group for International Agricultural Research* (CGIAR) and three additional advanced research institutions (non-CGIAR consortium members)

[j]http://www.icarda.org/cac/

The *International Fund for Saving the Aral Sea* with the five Central Asian countries as member states, coordinating cooperation at national and international levels in order to use existing water resources more efficiently and to improve the environmental and socioeconomic situation in the Aral Sea Basin

Or the *Central Asia and Caucasus Association of Agricultural Research Institutions*,[k] which aims at facilitating regional cooperation in agricultural research by providing a dialogue platform to the various stakeholders of the agricultural arena and by supporting information flow from globally operating organisations to local partners and back

[k]http://www.cacaari.org/

In the 1990s, the European Union launched *Partnership and Cooperation Agreements* with the Central Asian countries which also provide an umbrella for cooperation in S&T. Since the adoption of 'The EU and Central Asia: Strategy for a New Partnership' by the European Council in June 2007, the EU has intensified its relationship with the whole region. The strategy is supported by a significant increase of the EU's technical assistance in the region supporting higher education cooperation and academic and student exchanges under the new Erasmus Mundus facility and TEMPUS.

The Development Cooperation Instrument (DCI) (2007 to 2013) is a European programme for poverty reduction, sustainable economic and social development and the integration of Central Asia into the world economy. It is endowed with a total budget of 719 million. In general, DCI projects are not targeting dedicated R&D topics, but some of the projects include scientific knowledge generation activities and are therefore - at least to some extent - related to scientific research. Out of 176 supported projects, 29 contain educational and scientific issues. Nevertheless, there seems to be a lot of room for advancing the link between scientific research and problem-solving approaches for poverty reduction and social and economic development.

Summing up the EU perspective towards R&D cooperation with EECA countries, the CREST working group on R&D internationalisation revealed that almost all European countries rank USA, China and Japan as the most important target countries for cooperation within their own R&D internationalisation policy focus. Russia and India were among the next group of target countries, but mentioned by a significantly lower number of EU member states. With the exception of Ukraine, all other EECA countries were not among the prioritised cooperation target countries (Sonnenburg et al. [2008]). On the other hand, some Eastern European countries are in the immediate neighbourhood to the EU bordering Norway, Finland, Estonia, Latvia, Lithuania, Poland, Slovak Republic, Hungary, Romania and via the Black Sea also Bulgaria. Moreover, as described above, cooperation with the European neighbourhood countries, also in the field of S&T, is an explicit priority of the EU's foreign policy.

# Current State of S&T in the Eastern European and Central Asian Countries

The R&D capacities of the EECA region, are characterised by a general low level of R&D expenditures (except for Russia) which generate only limited scientific and economic results. Funding for R&D in the five Central Asian countries is generally low and ranges from 0.06% of gross domestic product (GDP) (Tajikistan) to 0.21% (Kyrgyzstan) in 2011 (see Table 1). Also, in most Eastern European countries, R&D Gross expenditure by GDP is very low. Yet, three groups can be differentiated: The highest values are observed in Belarus and Ukraine, with an R&D expenditure rate of 0.65% and 082 %, respectively (Table 1). The second group - comprising Georgia and Moldova - spends around 0.4% of their GDP on R&D. Lowest R&D expenditure was reported for Armenia and Azerbaijan with less than 0.3%, which is similar to R&D spending in the Central Asian countries. However, positive trends can be observed too. In some cases, the change might seem undetectable, e.g. in Belarus where the expenditure share remained almost unchanged in the period from 2001 to 2009, but since the country's GDP rose very rapidly, the amount of funding in nominal terms has also increased. In some cases, the goals to improve the situation are ambitious; such is the case in Azerbaijan where a recently announced strategy for S&T foresees a tremendous increase from 0.2% to 2% by 2015. However, it is also true that in some cases, the spending dropped drastically as a result of the recent financial crisis.

**Table 1:** Main S&T indicators of the EECA countries

| EECA countries | | R&D expenditure as % of GDP (GERD) | Number of research organizations | Number of R&D personnel |
|---|---|---|---|---|
| Central Asian countries | Kazakhstan[a] | 0.16 | 424 | 17,021 |
| | Kyrgyzstan[b] | 0.21 | 84 | 5,125 |
| | Tajikistan[c] | 0.06 | 67 | 5,617 |
| | Turkmenistan | n/a | 46[h] | 3,689[j] |
| | Uzbekistan[d] | 0.20 | 202 | 34,587 |
| Eastern European countries | Armenia | 0.27 | 83 | 6,926[k] |
| | Azerbaijan | 0.2 | 146 | 22,500 |
| | Belarus | 0.65[f] | 446 | 20,571 |
| | Georgia[e] | 0.4 | 31 | 3,200 |
| | Moldova | 0.42 | 38 | 4,764[l] |
| | Russia | 1.24 | 3,536 | 742,433[m] |
| | Ukraine | 0.82[g] | 1,303[i] | 141,000[n] |

[a]According to the Kazakhstan Agency for Statistics. http://www.stat.kz; [b]National Statistic Committee of the Kyrgyz Republic, 2010; [c]UNESCO Science Report 2010; [d]Committee for Coordination of Science and Technology Development of Uzbekistan 2010; [e]Source: SRNSF; [f]Science, Innovation and Technology in the Republic of Belarus – 2008. Statistical book, State Committee on Science and Technology, Ministry of Statistics and Analysis of Belarus, 2009; [g]State Statistics Service of Ukraine: Science and Technology Activities in Ukraine - Statistical Data Collection (Державна Служба Статистики України: Наукова та інноваційна діяльність в Україні - Статистичний збірник, ДП 'Інформаційно-видавничий центр Держстату України') Kiev, 2011, p. 178 (data for 2010)[;] [h]Estimated; [i]State Statistics Service of Ukraine: Science and Technology Activities in Ukraine - Statistical Data Collection (Державна Служба Статистики України: Наукова та інноваційна діяльність в Україні - Статистичний збірник, ДП 'Інформаційно-видавничий центр Держстату України') Kiev, 2011, p. 10 (data for 2010); [j]Statistical Yearbook of Turkmenistan, Ashgabat, 2010, p.160; [k]National Statistical Service of RA, http://armstat.am/ (data for 2009); [l]The Court of Accounts of Moldova Report, http://lex.justice.md/viewdoc.php?action=view&view=doc&id=338497&lang=1; [m]Number of researchers is 369, 237 (2009); [n]State Statistics Service of Ukraine: Science and Technology Activities in Ukraine - Statistical Data Collection (Державна Служба Статистики України: Наукова та інноваційна діяльність в Україні - Статистичний збірник, ДП 'Інформаційно-видавничий центр Держстату України') Kiev, 2011, p. 31 (data for 2010); number of researchers is 89,600.

Schuch *et al.*

Schuch *et al. Journal of Innovation and Entrepreneurship* 2012 1:3 doi: 10.1186/2192-5372-1-3

In the EECA region, Russia presents the highest R&D quota (1.24% in 2009), although it grew even further in 2010. Russian R&D allocation in 2008, expressed in purchasing power parity, corresponded roughly to the R&D allocations of Canada, India or Italy (HSE [2010]). In Russia, like in the other EECA countries, R&D is largely funded from the state budget, and the scarce resources are mainly concentrated in the public-influenced sector, usually characterised by low research commercialisation results.

Typically, state-funded R&D is allocated through core funding and/or through competitive mechanisms such as programme type schemes and competitive grants. In certain countries, however, (e.g. in Belarus) the predominant method for financing research has the characteristics of public procurement, with the project proposals selected on a competitive basis, either for basic or applied research, and the results owned by the state or state-owned organisations.

Like other EECA countries Russia also faced a significant decrease of the number of researchers.[i]R&D personnel in the Russian Federation counted 742,433 heads in 2009, which is 2/3 of the 1991 value. In full-time equivalents, Russia has five times more R&D personnel employed than Brazil, Canada or Italy and little less than Japan. The percentage of R&D personnel by 10,000 employees brings Russia at equal level to Germany and above Korea or the United Kingdom. However, only half of the R&D personnel in Russia are researchers, and therefore, in reality, Russia clearly falls behind Korea and UK. Since 1991, the highest drop in absolute numbers of researchers occurred in the business-enterprise sector, which was the largest employer for researchers in the country.

[1]An exception is Belarus, where R&D employment increased by 5% between 2003 and 2008.

Ageing of R&D personnel remains a problem in most Eastern European countries. Since a considerable part of the most active mid-age and young scientists have moved abroad or left the research sector, the research teams are currently composed to a large extent by researchers close to the retirement age. In Russia, more than 50% of researchers are above 50 years of age. Low wages and weak career prospects for young researchers are a common issue, resulting in a continuous brain drain problem. However, attempts are made to attract young scientists, usually through involvement in international programmes and/or through incentives to the diaspora (e.g. in Armenia and Moldova, and recently initiated in Georgia).

Regarding the number of research organisations in absolute figures, Russia and Ukraine have, by far, the highest numbers followed by Belarus, Kazakhstan, Uzbekistan and Azerbaijan (see Table1). All the other EECA countries count less than 100 research organisations. In most EECA countries, the national Academy of Science is central to the research system of the country. In some countries, like in Moldova, Tajikistan or Kyrgyzstan, it even defines and coordinates the research activities across all public institutions, including public universities, and manages state funds for basic research (the latter also in Ukraine and to a certain extent, also in Russia). In Turkmenistan, the vice prime minister for science, new technologies and innovation is also the president of the Academy of Science. The president of the Moldovan Academy of Science has a similar influential position. In almost all EECA countries, the academies in the future are supposed to be more engaged in applied research and in the cooperation with universities and the economic sector. For instance, in 2011, Armenia adopted a new 'Law on the National Academy of Sciences of Armenia' which stipulates wider possibilities for the Academy to carry out business activities and to commercialise R&D results.

With few exceptions, like in Georgia where the research institutes have been integrated in the university system, the universities until recently have occupied a rather modest place in the EECA research systems. Only approximately 40% of the 1,114 higher education institutes in Russia (data for 2009) are actually involved in R&D, and only approximately 20% of all professors and teachers conduct research (I Dezhina and M Spiesberger, unpublished work). Nonetheless, the situation is changing. Funding from the academies is increasingly redirected to universities through a number of new initiatives, such as the awarding of a special status of 'Federal University' or 'National Research University'. These statuses are accompanied by generous federal budget funding. Ukraine also attempts to boost the integration of research in universities ('Programme for Science in Universities 2008-2012'). Moreover, in Ukraine, Belarus, Georgia and Armenia, the reform of the higher education system along the lines of the Bologna process is a priority.

Although almost all EECA countries aimed to reform their S&T systems in a way to respond to new economic and social requirements by introducing previously non-existent mechanisms (e.g. introduction of competitive funding schemes; enhancing linkages with universities and teaching), their national systems of innovation still have striking weaknesses in interlinking with economic and societal demands as well as with different fields of policy. Despite the intention of the political elite to consider innovation-oriented R&D agendas a priority and to support a diversification of the economy beyond primary goods production, R&D performance in the business-enterprise sector is still weak. Even in Russia, the number of small innovative enterprises is remarkably limited and estimated at 25,000. It should be noted, however, that some statistical appropriation problems hinder an exact assessment. In general, it can be concluded that SMEs are still not in a position to act as engines of innovation and that large enterprises account for the majority of innovation activities.

# The White Paper Recommendations in the Light of International S&T Cooperation Policy Objectives

A group of organisations from nine EU member states, two countries associated to FP7 and nine EECA countries (including Moldova which became associated to FP7 on 1 January 2012), all of them with a public or semi-public mandate, working together under a joint international S&T policy coordination project (INCO-NET EECA), funded by the European Commission, prepared the 'White Paper on Opportunities and Challenges in View of Enhancing the EU Cooperation with Eastern Europe, Central Asia and South Caucasus in Science, Research and Innovation'. The White Paper is based on the conclusions of three policy stakeholder conferences organised under the INCO-NET EECA project (Athens/2009, Moscow/2010, Astana/2011), on fact-finding missions to all EECA countries, on a series of expert workshops on several relevant topics (e.g. S&T statistics) and on a public consolidation via the internet. Furthermore, the White Paper integrates extensive desk research and has been consolidated in a dedicated policy stakeholder conference in Warsaw (November 2011).[m]

[m] Activities organised in the context of the following projects funded by the European Commission (FP7) and dedicated to the support of the EU-EECA policy dialogue have been taken into account too:

'S&T International Cooperation Network for Eastern European and Central Asia – INCO-NET EECA'

'S&T International Cooperation Network for Central Asia and South Caucasus – INCO-NET CA/SC'

'Enhancing the bilateral S&T Partnership with the Russian Federation (BILAT-RUS)'

'Enhancing the bilateral S&T Partnership with Ukraine (BILAT-UKR)'

'Linking Russia to the ERA: Coordination of MS/AC S&T Programmes towards and with Russia (ERA-NET RUS)'

'Networking on Science and Technology in the Black Sea Region (BS-ERA.NET)'

The White Paper suggests 39 recommendations to improve R&D cooperation between the EU and the EECA countries, which is, on one hand, an indication for comprehensive deliberations but, on the other hand, also an indication for a lack of coordinated priority setting. Although a number of recommendations extend into the autonomous competences of state authorities and research performing organisations, the authors explicitly state that the 'White Paper does not intend to interfere with autonomous decision-making processes but to contribute to the knowledge base of the international STI cooperation between EU and EECA countries with an informed input that takes into account the international perspectives of different regions and countries' (Sonnenburg et al.[2012], p. 12). In this sense, the White Paper is less of a formal policy paper than an academic piece of work reflecting and deepening an ongoing policy dialogue, which is empirically evidenced. As a first step for a more consolidated policy approach, the White Paper consequently suggests the elaboration of a medium-term joint roadmap for enhanced science, technology and innovation (STI) cooperation between the EU, its member states and the EECA countries to be built on common goals for mutual benefit and to be implemented in partnership through joint instruments. It explicitly addresses the 'European Strategic Forum for International Cooperation' (SFIC) as the highest S&T internationalisation policy forum at the European level involving the EU member states, to launch a new SFIC pilot activity, by inviting EECA partner countries to join the dialogue and to monitor upcoming joint activities. With this suggestion for a procedural hand over, the perceived lack of policy legitimisation, caused by the facts that the White Paper is (1) a project deliverable and not an institutionalised spring-off and (2) prepared by a limited number of institutions (although by all means influential) which can neither represent the EU nor each single EECA country, should be overcome. The authors of the White Paper also call for more and broader consolidation when they write that the 'process of developing a joint roadmap needs to include wider stakeholder consultations in particular with the science community and the private sector in both regions' (Sonnenburg et al. [2012], p. 12).

The core part of the White Paper consists of the 'challenges and recommendation' section. The recommendations are disaggregated in five major topics, which are reflected in the context of S&T internationalisation objectives as outlined in the 'Internationalisation of R&D from an S&T policy perspective' section of this article, as follows.

# Adjusting and Implementing Policy Strategies

In this first thematic block, several recommendations directly related to strategic policy making and good governance are subsumed, such as generating, accessing and using standardised and internationally comparable data and knowledge for evidence-based policy making; embedding science, technology and innovation (STI) policy and policy delivery in a broader and aligned strategic policy system; building appropriate and internationally compatible national legal and ethical frameworks; strengthening the institutional fabric of the STI policy delivery systems with efficient tools and instruments; securing a sufficient financial allocation to the STI sector; identifying and addressing global and societal challenges, and making optimum use of international cooperation.

From an international R&D cooperation perspective, the focus of these recommendations is clearly on supporting strategic STI policy making, primarily perceived as a domestic endeavour, by implementing - as a supplement to the domestic homework - a series of international learning exercises (involving country representatives from the EU and EECA countries), improving existing international STI cooperation frameworks at national level (again based on international best practices exchange and/or international benchmarking exercises) and contributing to exchange and coordination activities at the international and global level. In this understanding, this first major block of recommendations is driven by an advanced interpretation of science diplomacy, which aims at supporting the establishment of a smooth multifaceted framework for future enhanced international R&D cooperation. The addressed actor is clearly S&T policy (respectively S&T policy makers), which could make use of experts as facilitators, who are either internal or external to the S&T policy system. While the actors targeted by these recommendations are from the sphere of policy making and policy delivery, the intended final beneficiaries of their work are, however, both private and public R&D organisations.

# Strengthening Research Performing Institutions

The second block of recommendations aims to strengthen research-performing institutions in order to make objectives related to international S&T cooperation attainable (e.g. tackling global challenges). In other words, research performing institutions have to be in the material and immaterial position to efficiently perform their duties (also in international division of labour), to adjust to changing demands of the society and economy and to possess capacities and capabilities needed for international S&T competition and cooperation.

It goes without saying that such a focus, directly targeting research-performing actors, is typically a central matter of domestic S&T policy. The White Paper complements this domestic perspective again by an international R&D cooperation perspective, suggesting to strengthen research-performing institutions through their involvement in international benchmarking exercises and twinning activities, which contribute to the adoption of good practices; to strengthen their strategic and operational capabilities through participation in international trainings and through the application of modern management techniques; and to establish and implement roadmaps, investment plans and management concepts for an improved development and exploitation of research infrastructures. Such an R&D internationalisation perspective falls partly under the development objective, where the improvement of local R&D capacities in developing countries recently gained attention (Bucar [2010]; Schuch [2007]), and partly under the cost optimisation objective when it comes down to a shared development and exploitation of research infrastructures.

Again, the actors addressed by this block of recommendations are largely stakeholders representing S&T policy making and policy delivery. Also, directors of R&D organisations, which are presumably public by nature, are directly targeted. It is likely that the interventionist power of public S&T policy will hardly reach out to private R&D organisations when strategic management issues (such as benchmarking, application of Balanced Score Cards or market foresight) are concerned. The final beneficiaries are again R&D-performing organisations.

# Strengthening Human Resources

The third block of recommendations identifies human capacity building as a particular challenge for all countries, especially in front of societal and economic transformation processes which require also an improved quality of communicating science to society. From the perspective of an international S&T cooperation policy, the adjustment of frameworks for international mobility is a particular challenge, especially in front of ongoing brain drain/brain gain dynamics which also take place between EU and EECA countries, expressing an unidimensional direction to the advantage of the more central EU and to the disadvantage of the more peripheral EECA countries.

The suggested recommendations, however, do not aim to put a firm halt on this lopsided brain movement, but rather follow a more liberal and multifaceted approach. The White Paper suggests to set up joint training and twinning activities, especially targeting young researchers; to further align scientific education schemes based on Bologna principles; to establish instruments for a more balanced mobility for students and researchers, e.g. through regional doctoral programmes; to further facilitate the issuing of scientific visa; to implement an EU-EECA Year of Science and to promote modern science communication.

From an R&D internationalisation policy perspective, the focus is on the resource acquisition objective, which in principle is true for both sides. Strategically important in this respect is the promotion and extension of European higher education standards (Bologna principles) towards the EECA region, which doubtlessly facilitates the international mobility of students and researchers, however, presumably at an uneven level-playing field which usually benefits the standard setter against the standard adopter. The actor targeted by these recommendations is primarily again S&T policy making and policy delivery, but a few recommendations can also be directly taken up by research organisations themselves. The intended final beneficiaries are, first of all, mobile (which typically means young) researchers.

# Strengthening the Role of the Private Sector

The fourth block of recommendations addresses R&D activities in the private sector. This is a policy issue which typically falls again under domestic S&T policies. From an international R&D perspective, the White Paper suggests some flanking measures, carried out through international cooperation, to support private R&D stimulation, e.g. through implementing joint international training courses on innovation management as well as international learning activities on stimulating the creation and nurturing of innovative companies and framework setting for a higher private engagement in science, technology and innovation; to provide linkages between industry-related R&D initiatives in the EU and similar structures in EECA and to establish joint competitive innovation funding programmes; to improve the conditions for investments in innovation and to encourage EU-EECA cooperation in this respect.

Most of these recommendations clearly fall under the market and competition objective, although often dressed in the wording of science diplomacy. Soft measures like international learning and exchange platforms, but also a few hard measures, such as establishing (and budgeting) competitive multilateral RTDI funding programmes, are explicitly suggested. Moreover, from an international R&D perspective, also policy issues are addressed which are increasingly dealt with at supra-national and/or global level (e.g. investment regulations at WTO level).

The addressed actor's level is that of S&T policy makers. The intended final beneficiaries are companies which should make use of the improved framework conditions and jointly established international instruments (e.g. joint calls for proposals).

# Strengthening Subregional Cooperation

The fifth block of recommendations focuses on the reduction of the fragmentation of the EECA region and on the increase of critical mass through subregional cooperation. In this respect, the White Paper suggests to strengthen subregional policy coordination and to stimulate networking between the science, technology and innovation communities, as well as to investigate the possibility of establishing regional centres of excellence. This block of recommendations is the least elaborated one because it refers only to the EECA region and does not emphasise international S&T cooperation between EU and EECA countries. It is a science diplomacy issue internal to the EECA region and clearly addresses the policy level as actor.

## Implementation Scenario

The White Paper suggests a short-term implementation scenario summarizing some of the suggested recommendations addressed to specific groups of stakeholders which can be implemented by utilising the existing cooperation instruments. Particular attention is given to existing programmes like the EU Framework Programme for RTD, the ENPI and the DCI as well as to ongoing and planned projects implemented hereunder such as the INCO-NET, BILAT and ERA-NET schemes. The White Paper also includes a roadmap, which suggests recommendations to be implemented at short, medium or long term, as well as a qualitative impact assessment, which estimates the expected impact of each recommendation on the national research performance, on human resources for R&D, on the innovation potential, on the participation in FP7 respectively its follow-up programme 'Horizon 2020', on addressing global challenges, on employment and on growth in general.

# CONCLUSIONS

By identifying the policy objectives addressed in the White Paper, it becomes obvious that the objective which is usually regarded as the major intrinsic objective for conducting international R&D cooperation (Boekholt et al. [2009]; Sonnenburg et al. [2008]), namely, the 'quality acceleration and excellence objective' is hardly directly addressed by the recommendations. Several suggestions, however, clearly aim to improve the state-of-art policies and instruments as well as the performance of research (organisations). These seem to be influenced rather by the spirit of science diplomacy and development cooperation than by the spirit of R&D excellence. The reason for this might be a perceived lopsided learning direction, which in general seems to go from the EU to the EECA countries (or possibly also from Russia to other EECA countries). This tilted position is neither a *priori* hegemonic nor surprising because many policy makers from EECA countries expressed the need to upgrade their national innovation systems, and the EU clearly offers a level-playing dialogue field for this with the INCO-NET, ERA-NET and BILAT-projects within FP7, which are used to establish platforms for learning and exchange. Presumably, with the exception of Russia, which progressed already a lot in modernising its national system of innovation (although without being in the position to lean back because S&T policy - like any policy field - is a moving target), most EECA countries are still undergoing basic reforms of their S&T system or have just implemented them. Their national innovation systems also suffer from a legacy of unfinished reforms. In most countries, amendments to laws, new strategy papers and even new institutionalised missions are more frequently published than rightfully implemented. Moreover, most countries of the region under scrutiny belong to laggards in terms of R&D intensity as shown in the 'Current state of S&T in the Eastern European and Central Asian countries' section.

Although the intended beneficiaries of the suggested recommendations featured in the White Paper are researchers and/or research organisations, both from industrial and academic background, the actor usually directly targeted by most recommendations is the S&T policy (delivery) system. The recommendations proposed in the White Paper feature an international cooperation dimension, which is understood to accompany domestic policies through offering relevant international dialogue, exchange and learning platforms. Most of the suggested recommendations are soft by nature, building on mutual interest, trust, benevolent interaction and a voluntarily participation based on 'variable geometry'. Such platforms, which also employ specific tools (e.g. international benchmarking exercises, S&T policy reviews, etc.) are considered as starting points for a subsequent improvement of the overall framework conditions for international R&D cooperation between researchers and research organisations from the EU and EECA countries. In such a step-by-step approach, also the ultimate quality acceleration and excellence objective is prepared and serviced, although other S&T policy objectives play a more immediate role in the short run, especially science diplomacy in its enabling function.

In that respect, the White Paper constitutes a valuable reflexive tool, and it can be expected that the implementation of the suggested recommendations will be beneficial not only to the EU-EECA cooperation, but also to the strengthening of the research systems in the EECA countries. However, to commence the implementation, it is necessary to identify and prioritize a limited set of recommendations. Such prioritization process can be the result of the continuing policy dialogue but would doubtlessly benefit from support at the highest political level at a certain moment (e.g. through the adoption of an action plan at ministerial level).

# AUTHORS' CONTRIBUTIONS

All authors read and approved the final manuscript.

# ACKNOWLEDGEMENTS

This article was prepared under the INCO-NET EECA project, funded by the European Commision. The authors acknowledge Tigran Arzumanyan, Adalat Hasanov, Olga Meerovskaya, Theodore Dolidze, Nikoloz Bakradze, Kamila and Sulushash Magzieva, Jyldyz Bakashova, Sergiu Porcescu, Diana Grozav, Anna Pikalova, Liliana Proskuryakova, Ilkolm Mirsaidov, Dovlet Jumakuliev, Vadym Yashenkov, Olena Koval, Rustam Saidov and Durdona Komilova for their contributions to the acquisition and interpretation of data.

# REFERENCES

1.   Archibugi, D, & Michie, J (1995). The Globalisation of Technology: a new taxonomy. *Cambridge Journal of Economics*, *19*(1), 121–140.

2.   Boekholt, P, Cunningham, P, Edler, J, Flanagan, K (2009). Drivers of international collaboration in research. Synthesis report to the EU Commission. Technopolis and Manchester Institute of Innovation Research, Brussels. April 2009

3.   Bucar, M (2010). Science and technology for development. Coherence of the common EU R&D policy with development policy objectives. Discussion paper 19/2010 of the German Development Institute. Deutsches Institut für Entewicklungspolitik, Bonn.

4.   Cantwell, J (1995). The Globalisation of technology: what remains of the product cycle model? *Cambridge Journal of Economics*, *19*, 155–174.

5.     Dalton, DH, & Serapio, MG (1999). Globalizing industrial research and development. U.S. Department of Commerce, Technology Administration, Office of Technology Policy, Washington, D.C.

6.     Edler, J (2007). Internationalisierung der deutschen Forschungs- und Wissenschaftslandschaft (Internationalisation of the German research and science landscape). Fraunhofer IRB Verlag, Stuttgart.

7.     Edler, J (2008). Creative internationalization: widening the perspectives on analysis and policy regarding beneficial international R&D activities. *The Journal of Technology Transfer*, *2008*(4), 337–352.

8.     Edler, J, & Boeckholt, P (2001). National public policies to exploit international science and industrial research. A synopsis of current developments. *Science and Public Policy*, *28*(4), 313–322.

9.     Edler, J, & Flanagan, K (2009). Drivers of policies for STI collaboration and related indicators. Literature Review, Manchester/Brussels. April 2009

10.    Edler, J, Meyer-Krahmer, F, Reger, G (2002). Changes in the strategic mangement of technology – results of a global benchmarking study. *R&D Management*, *32*(2), 149–164.

11.    European Commission (1999). Wrap-up INCO 1 (Final): statistical overview on international RTD cooperation in FP4 (1995–1998). European Commission, Brussels. July 1999

12.    European Commission (2000). EU co-operation with the NIS in science and technology. Office for Official Publications of the European Communities, Luxembourg.

13.    European Commission (2007). The European research area: new perspectives. Green paper presented by the Commission (SEC(2007)412, COM (207) 161 final. European Commission, Brussels.

14.    European Commission (2008). A strategic European framework for international science and technology cooperation. Communication from the Commission to the Council and the European Parliament. Brussels, European Commission.

15.    Godin, B (2001). What's so difficult about international statistics? UNESCO and the measurment of scientific and technological acitivities. Working Paper No. 13 of the Project on the History

and Sociology of S&T Statistics. Canadian Science and Innovation Indicators Consortium (CSIIC), Montreal.

16. Gokhberg, L, Gorodnikova, N, Wudtke, J (1999). Volkswirtschaft im Übergang. Ein Vergleich der F&E-Indikatoren in Russland und Deutschland (National Economy n Transition. A comparison of R&D indicators in Russia and Germany). Materialien zur Wissenschaftsstatistik, Heft 11, April 1999. Wissenschaftsstatistik GmbH im Stifterverband für die Deutsche Wissenschaft, Essen.

17. HSE (2010). Science and technology. National Research University - Higher School of Economics, Innovation. Information society. Pocket Data Book. Moscow.

18. Idenburg, P, Stalnacke, P, Schuch, K, Nyiri, L, Ventskonvsky, O, Mandrillon, M-H, Sokolov, A, Eikenberg, H, Sorensen, OJ (2004). External evaluation report on the programme of the International Association for the Promotion of Cooperation with Scientists from the New Independent States of the Former Soviet Union (INTAS) covering the period 1993–2003. INTAS, Brussels.

19. INTAS (1995). Revised statutes adopted on 14 November 1995 in Brussels: INTAS Secretariat. INTAS, Brussels.

20. Kougiou, S, Spiesberger, M, Kerasioti, V (2010). State of the art and perspectives of bilateral S&T programmes between EU MS/AC and Russia and of activities of S&T Programme Owners in EU MS/AC towards Russia and in Russia towards EU MS/AC accompanying / complementing bilateral S&T agreements. Deliverable 1.3 of the ERA-NET RUS project. [http://www.eranet-rus.eu/en/107.php.]Accessed 12 April 2012

21. Le Gohebel, M, Pecarz, D, Handler, K, Schuch, K (2011). S&T cooperation between the EU and Ukraine: benefits and barriers. *Foresight*, 5(3), 44–57.

22. Narula, R, & Zanfei, A In Fagerberg J, Mowery DC, Nelson RR (Eds.) (2004). Globalization of innovation: the role of multinational enterprises. *The Oxford Handbook of Innovation* (pp. 318–345). University Press, Oxford.

23. OECD (2005). Forum on the internationalisation of R&D. background paper: internationalisation of R&D – trends, issues and implication for S&T policies. OECD, Paris.

24. OECD (2008). Internationalisation of business R&D: evidence, impact and implications. OECD, Paris.

25. Rama, R (2008). Foreign investment innovation: a review of selected policies. *The Journal of Technology Transfer, 2008*(4), 353–363.

26. Sachwald, F (2008). Location choices within global innovation networks: the case of Europe. *The Journal of Technology Transfer, 2008*(4), 364–378.

27. Schuch, K (2005). The integration of Central Europe into the European system of research. Wien und Mülheim a.d. Ruhr, Guthmann-Peterson.

28. Schuch, K In Nechifor J, Radosevic S (Eds.) (2007). Towards the introduction of the system of innovation concept in official development assistance. *Proceedings of the International Conference and High Level Round Table "Why invest in science in South Eastern Europe"* (pp. 147–162). UNESCO BRESCE, Venice.

29. Schuch, K (2011). Indikatoren zur Messung der Internationalisierung von Wissenschaft und Forschung (Indicators to measure internationalisation of science and research). Final Report prepared for the Austrian Ministry of Science and Research, Wien.

30. Sonnenburg, J, Bonas, G, Schuch, K (2012). White paper on opportunities and challenges in view of enhancing the EU cooperation with Eastern Europe, Central Asia and South Caucasus in science, research and innovation. ICBSS, Athens.

31. Sonnenburg, J, Nill, J, Schuch, K, Schwaag-Serger, S, Teirlinck, P, Van der Zwan, A (2008). Policy approaches towards S&T cooperation with third countries. Analytical Report on behalf of the CREST Working Group on R&D Internationalisation, Brussels.

32. Spiesberger, M (2008). Country Report Russia. An analysis of EU-Russian cooperation in S&T. Prepared on behalf of the CREST OMC Working Group, Brussels.

33. Verbeek, A, & Shapira, P (2009). Analysis of R&D international funding flows and their impact on the research system in selected member states. RINDICATE report by IDEA Consult, NIFUSTEP and University of Manchester on behalf of the European Commission, Brussels. December 2009

# Chapter 6

# The Goal, Path, and Policy Responses of China's New Urbanization

Pengfei Ni[1]

[1]Center for City and Competitiveness, Chinese Academy of Social Sciences, Beijing, China

[2]Yuetanbeixiaojie, Xicheng District, Beijing 100836, China

## ABSTRACT

Urbanization has great significance for China's economic and social development, but the traditional model of urbanization is unsustainable. This paper details the basic model of new urbanization after analyzing briefly the requirements for implementing new urbanization. The basic model of new urbanization is, with the Scientific Outlook on Development as the guiding principle, to insist on comprehensive, coordinated, and sustainable development in building urban China. Urbanization of the population is the key content. Information, agricultural industrialization, and new industrialization are the driving forces. Productivity-induced economic growth is the development method. Government guidance and the market are guarantee mechanisms. The article lays out eight specific ways of integrating an integrated rural and urban China. It also proposes four basic responses to advance new urbanization: drawing up strategy and planning, providing infrastructure and public services, strengthening supervision and management, and improving institutions and policies.

# TIME CHARACTERISTICS OF NEW URBANIZATION

Urbanization, caused by scientific and technological progress as well as the development of productivity, is a historical process both of the agricultural population dispersing in agricultural functional areas and moving into the nonagricultural population in nonagricultural functional areas, as well as traditional rural society transforming into a modern urban society. The urbanized population is not only the traditional urban population of label significance, but is also a population that enjoys urban infrastructure and public service. The urban area is not only the area with administrative and geographic significance, but also the functional area of bearing non-agricultural population and industries. Chinese urbanization exhibits the increase of the number of cities, the expansion of urban areas, the increase of the urban population; as well as the change of jobs, the industrial structure, and spatial morphology. The great change of human society's organization methods, productive methods, and living method produces a new face of the economy, society, culture, environment, and human beings.

After 30 years of reform and opening, China has created a magnificent epic of urbanization to strongly support the miracle of economic growth. By 2011, the rate of Chinese urbanization had reached more than 50%, with an urban population of 670 million, 655 cities, and 20,000 small towns (Wang et al. 2012). The urban and town system is formed basically with different levels: large cities are centers, middle and small cities are the backbones, and small towns are fundamental. Today, Chinese urbanization is in the middle stage. The traditional urbanization mode, that is, economic development, is central target; export-oriented industrialization is central power; local government plays a key role; land is the main content; scale expansion is the development method; and the great volume of investment of material capital is the driving element. This is not sustainable. The traditional urbanization model has many problems, such as the population not being fully urbanized. Although 50% of the population lives in urban areas, only 35% of residents enjoy public services as residents with urban household registrations. Land is over-developed. Over the past 30 years, the population has doubled, while construction area has increased four times. Most people move into center cities with high administrative ranking, while middle cities develop slowly and small cities do not function well. Small towns are scattered. The traditional urbanization model causes other problems: high resource consumption, spatial centralization, imbalanced economic structure, pollution, and intensified social contradictions. Some cities and rural areas have serious 'urban disease' and 'rural disease', respectively.

Today, over 50% of the global population lives in the urban areas where science and technology changes rapidly, city development is more and more clustered, economic globalization is soaring, and the ecological environment continues to deteriorate. Economic globalization means that China may utilize all global factors, resources, and markets to push forward Chinese development. Information technology helps China digest and absorb global advanced technologies, such as advanced transportation technology and communications and networking technologies. Since world development is imbalanced, excessive consumption in developed countries and the rapid expansion of demand in emerging countries are causing a serious shortage of non-renewable resources. Meanwhile, excessive pollution in developed countries and increasing pollution in emerging countries is causing global warming and the continuing deterioration of the environment. For this reason, Chinese urbanization cannot take the traditional road.

China is now a middle-income country. Reform and opening has created a miracle of economic growth, which has enabled China to have enough capital, talent, and technology to transform and upgrade industries, including urbanization. Due to increasing individual, the enhancement of personal quality, social openness, and the awaking of citizen consciousness, China must look for new urbanization models to avoid the intensification of social contradictions and protect its fragile ecology. China is a large country with a large population and vast land, but resources are in serious shortage and distribution is imbalanced. There are great differences in economic, societal, and cultural development. Thus, Chinese urbanization must utilize its large-scale economic advantage to avoid over centralization and long-distance movement (Qiu 2003).

In summary, in this new era, it is important to reflect deeply on China's urbanization model, and fully explore the model, path, and strategy of new urbanization in China (Wang 2010).

# THE BASIC MODEL OF NEW URBAN-IZATION

On the basis of the problems of traditional urbanization, the time characteristics of future urbanization, and Chinese characteristics, this paper proposes a basic model of new urbanization: with the Scientific Outlook on Development as the guiding principle, to insist on comprehensive, coordinated, and sustainable development in building urban China. Urbanization of the population is the key content. Information, agricultural industrialization, and new industrialization are the driving force. Productivity-induced economic growth is the development method. Government guidance and the market are guarantee mechanisms.

## The New Final Target of Urban China is The Integration of Urban and Rural Areas

The final target of new urbanization is to build an urban China that integrates urban and rural areas. The integration of infrastructure will cover urban and rural areas in China, public services will be equal for everyone, and farmers and urban citizens will have the same opportunity to obtain knowledge, skills, personal quality, and income. The details can be described as following:

## New Economy

In the context of new industrialization, the industrial system consists of high-technology content, yields good economic returns, has low resource consumption, produces less environmental pollution, and is less labor intensive. The demand structure system is composed of domestic demand and consumer support. Human capital and technological innovation play a key role in the element structure system. Large companies and a great number of middle and small companies form the enterprise structure system.

## New Society

Under the administration of new community, new social organization, and new society, the communication method is social, the social connection is international, living methods are modern, and residents' composition is diversified and mobile. Social class structure shows an olive shape and is harmonious.

## New Environment

Beautiful natural scenery, good ecological environment, resource conservation, effective environmental protection, and a good relationship between people and the environment should be achieved.

## New Governance

For urban–rural development and urban life, provide a social environment with a good system and rule of law, an open, convenient, and free-market environment, a free and open cultural environment, and a democratic and free political environment.

# The New Main Content is Equal Public Services, Which is The Core of Urbanization

## Equal Access to Public Services among All Urban Residents Promotes People-First Urbanization

There are three kinds of population urbanization: university education urbanization, local urbanization, and migrant urbanization. Each portion makes up about 30%, while other urbanization accounts for 10%. The new main content of population urbanization is to give the three kinds of population urbanization equal access to public services, just like those with urban household registration. This is particularly necessary for farmer laborers. In the future, the main point of urbanization will be to provide equal public service for 0.16 billion farmer laborers and newly added farmer laborers.

## The Key is to Optimize the Utilization of Land to Promote Productivity-induced Economic Growth Urbanization

Land urbanization has mainly three fields: old urban area, new urban area, and rural settlement. The new content of land urbanization is to increase the land use efficiency by rebuilding old urban areas and to increase urban capacity by reducing the expansion of new urban areas. By adjusting rural settlements, the scale of rural construction land decreases. Simultaneously, the distribution of urban and rural areas should be optimized.

## It is Necessary to Overcome 'Urban and Rural Diseases' in Order to Keep Urbanization Healthy and Sustainable

After the rate of urbanization is over 50%, urban disease will burst centrally and rural areas may continue to deteriorate. Therefore, the main contain of new urbanization is to avoid 'urban and rural diseases' and to make sure urban and rural both win.

# New Basic Drivers are Information Technology, New Industrialization, and Modern Agriculture

China's new urbanization has three new drivers due to global information, key technological breakthroughs, and the third industrial revolution.

## *New Pulling Power*

Information technology is based on new industrialization with high technology content, which yield good economic returns, has low resource consumption, produces less environmental pollution, and is less labor intensive. Information technology and the intelligent third industrial revolution will form new pulling power for new urbanization.

## *New Pushing Power*

China should fully utilize modern information technology, agricultural production, and management based on information and modern service. Agricultural mechanization, agricultural science, agricultural industrialization, and labor quality enhancement will form a new pushing power for new urbanization.

## *New Source of Power*

Information technology is a main representative of modern science and technology that indirectly affects urbanization supply and demand and directly decides urbanization scale, speed and quality. Information technology, new industrialization, and modern agriculture provide more powerful sources of power than before and ensure that urbanization is people-first, driven by innovation and sustainable development.

# The New Development Method is Productivity-induced, Intensive Economic Growth

## *It is a Good Idea to Improve the Construction Density of Urban and Rural Areas and to Use Underground Space Fully*

It is necessary to increase industry and population density in limited urban space and to promote industrial clustering development and intensive land development.

### *To Optimize the Layout of Urban and Rural Areas*

Reasonable layout of urban and rural areas and convenient connecting channels can improve the efficiency of resources, factors of production, and the industrial arrangement.

### *To Improve the Function of Urban and Rural Areas*

Improving and building comprehensive urban and rural infrastructure and a public service system can improve service quality and attract investment.

### *Breeding and Utilizing High-end Factors of Production*

It is necessary to decrease material capital investment and resource consumption and to use human capital and innovation to promote urbanization.

# BASIC PATH OF CHINESE NEW URBANIZATION

The accomplishment of new-content urbanization depends on new driving forces and methods to carry out target urbanization. It is crucial to keep a general way of sustainable development and new urbanization. Sustainable urbanization is development balanced between the short-term and long-term and development coordinated among the economy, society, and the environment that will allow future generations to develop sustainably and live and work in peace and contentment.

## Walk the Road of People-first Urbanization, Pushing People to Move to Urban Areas from Rural Areas

To build city needs people and people will live in new urban areas; thus, all tasks of urbanization should focus on people. The core of people-first urbanization is to allow farmers to move to urban areas, while at the same time allowing farmers to change their jobs from farmers to urban residents. These farmers have the same rights as citizens in urban areas: housing, education, pensions, income, entertainment, and health care. They also have the right to enjoy modern urban infrastructure and public service, to enjoy a good residential space environment, beautiful ecological environment, and a clean and highly efficient production environment. The purpose of people-first urbanization is to promote comprehensive social development, to raise personal quality and incomes, to improve factor supply, to enlarge consumption, and to maintain sustainable development between urban and rural areas.

## Walk a Smooth Road of Urbanization Where Equality and Efficiency are Equally Important

Population concentration and free movement can ensure the best distribution of resources and shared-scale economy and external economy, so as to raise economic efficiency. However, population concentration will cause high costs, urban disease, and enlarging regional disparities. It is necessary to push forward new urbanization. First, the general layout of urbanization should be designed to form a unified national market to reduce obstacles among separated markets in different cities to take advantage of the scale economy of a large country. On the other hand, reasonable measures must be taken to optimize the concentration of urbanization and maximize the benefit of the scale economy and external economy. Second, regarding the relation of urban scale and urban development, the road of urban clustering development emphasizes the coordination of large, middle, and small cities. The road of central city driving small cities should be changed. According to the principle of marginal cost being equal to marginal revenue, cities of different sizes are built in the different regions based on position, transportation, resources, environment, factors, and market conditions. Infrastructure networks will connect all cities to form an urban ribbon and to complement each other, so that all cities will benefit from the external economy. Third, walk the road of integration between urban and rural areas, that is, the integration of industry and agriculture, urban and rural, and urban citizens and rural residents. The dual structure of urban and rural economies should be broken to promote urban and rural areas complementing each other and lasting prosperity. Industry feeds back into agriculture by transfer payments to support rural vulnerable groups.

## Walk the Road of Urbanization Where Industry and the City Interact to Ensure Lasting Prosperity

First, the advantage of secondary industry should be maintained or expanded. Raising the international competitiveness of secondary industry is crucial to providing jobs and increased incomes. Second, it is necessary to develop various businesses, including production, consumption, distribution, and social service, in order to promote economic growth and speed up urbanization. Third, agricultural industrialization should be accelerated to raise agricultural efficiency and farmer incomes. Fourth, the implementation of a resident income plan ensures rising incomes and more job opportunities to promote urbanization and economic growth.

# Walk A Green-Development Road of Urbanization and Ensure the Natural Beauty of the Ecological System

Future urbanization does not walk the road of both protecting the environment after pollution and protecting the environment first with economic development second. Regarding the Chinese development condition, both economic prosperity and protection of the environment are important to ensure coordinated and sustainable development between ecology and the economy. First, the energy-saving environmental protection industry should be encouraged, so as to reduce the consumption of energy, water, and air. Second, the circular economy should be emphasized to produce more and consume less. Third, industrial zones should be built so all enterprises may share external economy and wastes will be treated centrally. Fourth, the government should encourage urban residents to carry out energy-saving environmental protection in daily life and promote energy-saving environmental protection. Fifth, green areas should be increased to raise the self-purification ability of the environment. Sixth, green consumption should be encouraged to reduce consumption pollution. Seventh, occupation and allocation of resources should be managed reasonably to promote the positive interaction of urbanization and environmental protection.

# Walk the Road of Inclusive-growth Urbanization and Ensure Fairness and Justice

The relationship of urbanization and social development should be focused on to ensure harmonious urbanization. First, in the field of politics, people of all social classes should have more opportunities to discuss and decide their issues of concern. Government should provide a platform for all people to create their own careers and share urban progress results. Second, in the field of society, the government should treat the relationship of government, businesses, and residents reasonably, protect the legal rights of farmers and urban residents, and respects residents' willingness and choice. Government treats reasonably the relationship of local residents and migrants, protects legal rights of migrants and enable migrant have a good relation with local residents. Government should reasonably handle the relationship between different income groups, encourage people to innovate, create jobs and make money, and should simultaneously take care of and protect weak groups. Social insurance policies should cover weak groups and narrow the gap between rich and poor.

## Walk the Innovation-and-Drive Road of Urbanization and Ensure That Cities Lead the Future

As a latecomer to urbanization, China should not hesitate to carry out a strategy of innovation and drive. First, the implementation of a strategy of innovation and drive may push economic growth, and the latecomer may become the leader. Therefore, the priority development of education is a key step by continuing to increase education investment, to lengthen compulsory education, and to extend occupational education and adult education. A multi-level education system matching the future Chinese economy should be established. Increasing science and technology investment ensures breeding the innovation main body, creating the innovation network, setting up the innovation platform, and forming a science and technology innovation system that will match urbanization development and the industrial system. Second, the implementation of institution and management innovation will ensure a sustainable new urbanization. Third, innovation is not only in technology, but also in management, particularly in development methods. China should strive for latecomers to become leaders.

## Walk the Road of Government-guided Urbanization and Develop the Basic Role of the Market

Urbanization is the exhibition of main market entities sharing a preference for the external economy in space accumulation. It is also the accumulation and transformation of rural population to urban areas. Improving market system is conducive to the process of free selection of market entities. Because of the existence of market failures, it is difficult to achieve optimal equilibrium only by market selection. The government has to create suitable soft and hard conditions to promote healthy sustainable urbanization and development. On one hand, the flow of market entities helps to show their space preference. On the other hand, more attention should be paid to fairness or efficiency of land utilization. Therefore, the main functions of the government are to make forward-looking scientific planning, to build infrastructures in administrative areas, to provide equal public service for all residents, and to create fairness, justice, equality, and standard institutional conditions.

# Walk the Road of Open Urbanization and Ensure the Individual Character of Cities

The advancement of new urbanization should face the whole world and have a global view to construct a fully international country and urban–rural areas. The strategy is to open to the whole world; to use global elements, resources, and markets; to learn from global development experiences; to insist on global standards; and to follow global development trends. Then, the advancement of new urbanization should stand on the local land and have individual character. Each city has its own individual character due to its different fundamentals, background, and environmental and development conditions. Large and central cities should perform the gathering function to develop technology-intensive and capital-intensive industries, so as to enhance radiating driving capacity to their regions. Productive service industries should also be developed to enhance technological innovation and institutional innovation. Large and central cities should become main powers to participate global competitiveness. Medium and small cities should keep industrial characteristics and accept industrial transfer and strong support from large cities. Since investment cost is low and industrial development conditions are better in medium and small cities, manufacturing industry is the main content to raise effective power to surrounding areas. Finally, the integration of foreign culture and native culture is also important. The absorption of foreign culture is the same as all rivers run into the sea. All urban residents should carry out protection and inheritance of their cultural heritage, respect native cultures and lifestyle characteristics, and reflect the fusion of ancient and modern cultural charm. The coexistence and integration of the development of diverse cultures is a remarkable characteristic in new urbanization.

# POLICY RESPONSES TO ADVANCE CHINA'S NEW URBANIZATION

The basic model of new urbanization is, with the Scientific Outlook on Development as the guiding principle, to insist on comprehensive, coordinated, and sustainable development in building urban China. Urbanization of the population is the key content. Information, agricultural industrialization and new industrialization are the driving forces. Productivity-induced economic growth is the development method. Government guidance and the market are guarantee mechanisms. The article lays out eight specific ways of integrating an integrated rural and urban China. According to the functional principle of being perfect to make up for market failures, the government may implement four basic countermeasures to advance new urbanization actively and steadily: drawing up strategy and planning, providing infrastructure and public service, improving institutions and policy, and strengthening supervision and management.

## Formulate Strategy and Plans and Lead the Sustainable Development of New Urbanization

### The Development Goal is Integrating Urban and Rural Areas to Form an Urban China

The government should formulate an urban and rural development plan in 2030 and in 2050. Under the guidance of the Scientific Outlook on Development, top-level design should be made on the development goals and steps of Chinese urbanization and given full consideration.

The pace of urbanization should be kept in a reasonable range based on the ultimate loading condition of economic social development and the resource environment. In view of the global competition pattern and the future situation of industry transferring and upgrading, the annual growth of urbanization speed cannot and should not maintain the 1% annual growth of the past. Future Chinese urbanization should combine connotation urbanization and extension urbanization and focus on quality. It is necessary to solve the problem of the partial urbanization of 160 million people and to continue to fight to resolve the urbanization of 360 million new people.

## The Total Scale and Level of the Urban Population

The total population from 2030 to 2050 is estimated to stay at around 1.4 billion. The population space distribution will be 25% living in rural areas, 25% in small towns, 25% in medium cities, and 25% in metropolises (Jian and Huang2010). The urbanization level will be 75% on average (80% in the east, 70% in central regions, and 60% in the west). The total urban population will reach 860 million in 2030 and 1.06 billion in 2050.

## Construction Land and Level of Urbanization

The construction land for urbanization is 65% of non-farming construction land. Total non-farming construction land is 160,000 km$^2$; non-farming construction land in urban areas is 100,000 km$^2$, with large cities at 15%, medium and small cities at 20%, small towns at 30%, and rural areas at 35%. The standard of population density is 10,000/km$^2$.

## Urban Scale System

Cities are concentration points for national resources and fountainheads of economic development, as well as centers for national innovation and enrichment. The number of cities is around 1,000 and small towns 20,000 in 2030 and 2050. The urban space layout, which includes super cities, large cities, middle cities, small cities, small towns, and resident settlements, is rational and coordinated. First, the advanced techniques and methods of transportation and communication are used to develop a number of large and super large cities by rational planning. Both expanding large city scale and raising city quality are important to be growth poles and power sources. Second, a number of small cities, which are regional centers, are extended to medium cities to enhance driving power. The central government should support the development of medium cities. Third, the functions and management level of small cities should be improved. Fourth, small towns have special functions to connect urban areas with rural area. It is urgent to adjust the distribution of small towns.

## Urban Function System

In the view of urban functions, urban functions will tend toward multiple levels. Chinese cities will bean open system in the future: one global top city, 3 to 5 global cities, about 15 international cities, 30 to 50 national cities, and about 1,000 regional cities. Hong Kong, Shanghai, etc., will have concentrated high-end service industries and high-tech manufacturing. Research and design centers of industrial process and brand marketing will be developed dramatically to form a top city and a center of global economic control and management. Some cities being international cities will play professional functions to provide international services. Others being national cities or regional cities will play a comprehensive function to provide services and management for national or regional areas.

## Urban Space Layout

In the future, a lot of Chinese cities and towns will be distributed regularly to form urban clusters or urban ribbons, showing three structures: spot, line, and centralized area. The layout of the three structures will be connected with a modern transportation network. Cities will be the heart and vast rural areas the body.

First, in the national space layout: in eastern China, the construction of centralized areas is principle and combination of spots and lines is secondary. There are superior locations, a suitable environment, and dense population. It is easy to develop first, with transfer and upgrading, facing the world and participating in competition. In central and northeast China, the construction of lines is principle and combination of spots and centralized areas is secondary. The area is resource-rich and population is dense. The development of incentive policies will attract talent and industry to transfer here. The rise of central China and the revitalization of the northeast will be carried out. It is important to enhance conservation of resources and environment protection. In west China, there is with much land and few people. The construction of spots will be principle and combination of lines and centralized areas will be secondary. The implementation of the western development policy is ongoing. It is important to enhance environmental protection and ecological restoration.

Second, in the regional space layout: it will be crucial to take advantage of the scale economy and the external economy to push forward the integration of urban and rural areas and avoid malignant competition. Urban clusters will be formed with rational industrial distribution, favorable cooperation, a closely connected economy, as well as inside gathering and outside expanding functions. The aid of border areas is necessary.

Third, in the urban and rural space layout: clusters and networks of cities and towns will be built to benefit the integration of urban and rural areas, favorable cooperation, mutual benefit, win-win, and external economy. Compact cities will be built to reduce the spread and expansion of construction land. By building new rural areas and canceling and merging settlements, the new face of rural areas will appear and land will be used centrally to ensure farmland.

## Development Pace: Two Steps to Achieve the Ideal Goal of Urbanization

*First step* in 2020, the urbanization rate will be over 60% and the three-dimensional structure will change to a two-dimensional structure. The comparatively perfect urban/rural infrastructure will be basically constructed. In urban areas, basic equal public services will be provided to everyone and income will also be the same for everyone. In rural areas, infrastructure, public services, and income will be improved gradually.

*Second step*: in 2040, the urbanization rate will be over 75% and urban and rural two-dimensional structure will change to a one-dimensional structure. Perfect urban/rural infrastructure will have been built. Incomes and equal public services in rural areas will be the same as in urban areas.

# Provide Infrastructure and Public Services to Support the Healthy Development of New Urbanization

## *Infrastructure: Gradually Realize the Integration Urban and Rural Infrastructure and The National Integration Of Infrastructure*

### Build a Nationally Integrated Infrastructure Network

According to the layout of national urban and rural areas and the distribution of cities, national transportation, communications, and information infrastructure network systems, which are fast, high efficient, convenient, and low cost, are constructed and improved by the help of advanced science and technology, so that resources, people, and information can move freely among regions, cities, and urban and rural areas.

### Improve Regional Integration of the Infrastructure Network

Further accelerate the construction of urban physical infrastructure and intangible service networks so as to improve regional integration of infrastructure networks composed of highway, high-speed railways, channels, transportation pipelines, power transmission, networks, drainage pipe networks, and communication backbones, to decrease commuting distance among cities, and enable resource and information flow to realize the co-construction and co-sharing inside urban clusters and among large urban clusters.

Improve urban and rural integration of infrastructure networks. According to the plan for urban and rural integration, urban infrastructure networks will be extended to rural settlements and edge areas, and rural residents will have the same infrastructure network services as urban residents do.

## Public Services: Gradually Advance Equal Public Services in both Urban and Rural Areas and the Nation

On the basis of children education, social insurance, housing, and pension, in that order, farmer labors will gradually receive public services equal to those of urban residents. The employment and job creation system will be improved to raise resident incomes. The urban healthcare system will be improved to expand the coverage of healthcare insurance. The support capacity of affordable housing will be increased and public housing supply is enhanced. Simultaneously, it will be necessary to realize the integration and equality of employment, social insurance, education, healthcare, and other public services among regions and urban/rural areas.

Positive and efficient fiscal and tax, finance, industry will be formulated to aid rural areas, non-urban clusters, and edge areas. Large investment will be focused on infrastructure and public services to avoid the Matthew effect and to reduce the degree of regional tilt.

# Improve Supervision and Management and Promote the Smooth Development of New Urbanization

To realize efficient supervision and management for market entities, it will be necessary first to establish an administrative supervision and management system.

## Build an Administrative Management Framework Suited To an Urban China

Administrative level will be simplified and the city establishment model will be reformed. To adapt to administrative requirements, by learning from the experiences of advanced countries, the adjustment of administrative divisions and reduction of administrative level should be done quickly. There are 30 provinces, autonomous regions and municipalities as well as 14 cities specifically designated in the national plan. In east or central China, some developed regions with dense populations may set up new municipalities based on regional politics, economy, culture, and history. It is also possible to reduce provincial scale and allow provinces to manage counties directly.

## Build Longitudinal Intergovernmental Relationship of Property Power and Financial Power

To define property power and financial power of central and local governments is crucial for making clear the public responsibility and the public authority and making clear the property power and the public expenditure responsibility of all level governments. Since property power is consistent with financial power, the local government is very conscientious in the performance of its duties to promote urban development smoothly. At the same time, departments of the central government invest directly in some cities in poor regions by national projects or by fiscal transfers. The development of project grants from the central government is necessary and may guarantee the development of poor regions.

## To Build Up the Transverse Intergovernmental Regional Coordination Mechanism

The regional coordination mechanism will be improved further by setting up regional organization of urban development particularly inside urban clusters. Innovation in the model of public management is created by co-building market and infrastructure, co-coordinating regional industrial planning, and co-constructing transportation networks and information networks.

## Build Up the Integration of Urban and Rural Management Systems

First, the integration of urban and rural management systems led by cities is built to adapt to a complex economy, society, environment, and foreign connections. The urban management function is improved. Second, the idea of a public management is reformed to push forward the government management of humanization, efficiency, institutions, rewards and punishments, and competition, to raise public service efficiency and quality and to enhance urban competitiveness. Third, non-governmental organizations (NGOs) have developed widely, and a lot can be done by the NGO. Fourth, to enhance urban community construction, the city encourages organizational innovation and management innovation of community management to perform urban community function of politics, self-management, study, and service, to raise the level of self-education, and to encourage public participation. Residents can participate in managing the community and express their interest demands. Fifth, to enhance the urban emergency mechanism, cities should set up and improve urban emergency mechanisms, enhance crisis response capacity by institutionalized prevention and management system, and maximize the elimination of all kinds of unexpected events and hidden dangers or reduce the harm to a minimum.

## Improve Performance Evaluations and Accountability

Urban governments should establish a scientific performance appraisal system, formulate a performance evaluation index system, which is guided by the principle of the Scientific Outlook on Development, to show comprehensive, coordinated, and sustainable development. The evaluation index system should exhibit the standard of new urbanization, which includes urban economic development, social development, cultural prosperity, environmental protection, and so on. Evaluation results will reflect government performance and act as the key indicator in the job changes of main urban leaders.

## Redefine the Standards of City Establishment and Related Functions

First, revise the Urban–rural Planning Law. Re-clarify the scale and rank of cities on the basis of China's larger population. The standards of city establishment and scale by population size are super large cities, super cities, large cities, medium cities, small cities, and small towns. According to the urban scale, urban functions are decided. It is possible to merge and reduce some cities in western China to reduce management costs. In eastern China, some towns will be changed to cities to improve their functions. Second, formulate and revise relevant laws. By formulating a system of constraining, quantitative, clearly regulate urban resource utilization, environment protection, and the standards of infrastructure and public service based on urban scale.

# Reform of Institutions and Policy to Ensure the Healthy Development of New Urbanization

Establish Sustainable Institutions and Policy to Guarantee Urbanization

## Deepen Reform of the Fiscal and Taxation System

The central government should establish a longitudinal intergovernmental relationship of property power and financial power, carry out firstly compulsory education funds of national coordination and gradually implement national coordination of social insurance completely or in part. The central government should build up the central and local government fiscal transfer system to support farmers changing to urban residents. It is necessary to set up a cost-sharing system among governments, businesses, and individuals to help migrant workers change to urban residents and further to form a local fiscal and taxation system.

## Deepen Reform of the Land System

Establish an integrated urban–rural land system. Rural land ownership belongs to the nation, and farmers have land-use rights. By giving farmers transaction-use rights, there are improved rural land property rights to activate farmers' land assets by promoting land assets stock, farmer shareholders of land stock, and farmer democratic rights. The physical market of rural land property right transfer is justice, openness, and fairness. Land mortgage or sale is possible and land property rights protection system is established. Generally, land use, transaction, and income apportionment system are urban and rural integration, with the same right, same price, and free competition.

## Deepen Reform of the Household Registration System

Some migrants should become citizens based on conditions of employment, investment, and residency and for migrants who have not yet become registered urban residents, to establish a residential permit system for shared right and responsibilities.

## Deepen Reform of the Employment System

Further liberalize the labor market and open all industries to laborers. At the same time, cancel all local policies and regulations that limit rural laborers from moving to cities to ensure that rural labors have an equal right to employment. Cancel the discriminating employment system of the urban labor market to build up a unified employment training system.

## Improve the Social Insurance System

First, improve the system to reform social insurance system positively, integrate the urban and rural social insurance systems to change the division of urban and rural areas. Second, the width and breadth of social insurance system should be extended further, including national endowment insurance, unemployment insurance, healthcare insurance, injury insurance, and maternity insurance. Third, form a new social assistance system, in which minimum living security is fundamental, special assistance is supportive, and charity aid is supplemental. Fourth, local social insurance systems should have their own character.

## Deepen Reform of the Financing System

Under the guidance of the government fiscal, deepen the reform of the system by actively and innovatively develop direct financing tools mainly in the form of mid- and long-term credit support. Improve the finance system by enhancing institutions and market construction. Diversified financing models include infrastructure securities, infrastructure investment funds, municipal bonds, and private finance.

# Establish a Long-Term Mechanism to Ensure to the Continual Development of Urbanization

## Price Adjustment Mechanism

First, by advanced techniques and methods, make sure infrastructure and public services have excludability as much as possible. User-pay service employs gradient-type markup pricing, and use more and pay more, to reduce the burden on the poor population. Second, the market will determine the price of transfer of management right of rural land, the price of collective land transfer into state-owned land in designed construction areas and the price of urban removing and resettlement houses. The role of price adjustment in land use will be fully employed to realize land benefits allocated fairly and reasonably among farmers, urban residents, the government, and businesses in order to promote social harmony and improve land use efficiency.

## Ecology Compensation System

The price leverage effect is crucial in resource conservation and environment protection. It must clarify property rights and make users and polluters pay and ensure internalization of the external cost of economic production and daily life.

## Land Occupation Mechanism

Establish an index number of non-agricultural land area occupied per capita in different areas on the basis of the current population and land distribution. At the same time, make vouchers for population and land-use indicators. Each city has the right to decide on accepting a number of farmer laborers and a land-use index number in the light of its capacity. Cities where farmer laborers will move must provide houses, healthcare, and social insurance for farmer laborers and pay land-use index money. Rural areas, in line with population movement, will obtain land-use index money and part of the money will be used to transfer housing land into farmland.

## Fundraising Mechanism

Urban infrastructure construction needs large investment and the return cycle is long. There is a mismatch in the supply and demand of fund maturities. Therefore, develop long-term finance tools, such as long-term credit, fund investment, long-term funds, and asset securitization to match supply-and-demand cash flows for infrastructure construction.

## Fiscal Subsidy Mechanism

Central and local financial administrations should set up transfer payment subsidies for migrant laborers changing to urban residents in light of migrant worker residence size. Fiscal subsidies are mainly used to increase social insurance guarantees for migrant workers and their families, including medical facilities, compulsory education, and occupation education facilities, low-rent housing, and expansion of municipal facilities.

## Tax-adjustment Mechanism

Reform and improve the taxation system quickly. It is urgent to levy real estate taxes, land taxes, resource taxes, and environment taxes. The problem of 'land finance' must be resolved and the tax leverage effect will be fully employed in resource conservation and environment protection

# REFERENCES

1.    Jian X, Huang K (2010) Empirical analysis and forecast of the level and speed of urbanization in China . Econ Res J 3:28-39.

2.    Qiu B (2003) The characteristics and driving force of China's urbanization and its control by urban planning . Urban Studies 1:4-10.

3.    Wang G (2010) Urbanization: the core of China's economic development model transition . Econ Res J 12:70-81.

4.   Wang L, Li CC, Ying Q, et al. (2012) China's urban expansion from 1990 to 2010 determined by satellite remote sensing. Chin Sci Bull 57:2802-2812.

# Chapter 7

# Successful Education for AEC Professionals: Case Study of Applying Immersive Game-Like Virtual Reality Interfaces

Farzad Pour Rahimian[1], Tomasz Arciszewski[2], and Jack Steven Goulding[1]

[1]Centre for Sustainable Development, The Grenfell-Baines School of Architecture, Construction and Environment, University of Central Lancashire, Preston PR1 2HE, UK

[2]Civil, Environmental and Infrastructure Engineering Department, Volgenau School of Engineering, George Mason University, Fairfax Virginia, USA

# ABSTRACT

## Background

Global competition and the transdisciplinary nature of evolving Architecture-Engineering-Construction (AEC) activities makes it progressively important to educate new AEC professionals with appropriate skill sets. These skills include the ability and capability of not only developing routine projects, but also delivering novel design solutions and construction processes (some of which may be unknown), to feasible, surprising, or potentially patentable solutions. For example, despite recent innovations in immersive visualisation technologies and tele-presence decision-support toolkits, the AEC sector as a whole has not yet fully understood these technologies, nor embraced them as an enabler.

## Methods

Given this, this paper proposes a new approach for delivering education and training to address this shortcoming. This approach focuses on doing traditional (routine) work with creative thinking in order to address these challenges. This rationale is based on the principles of *Successful Education* as a new paradigm for engineering education, which is inspired by the Theory of *Successful Intelligence*, by the *Medici Effect* and Leonardo da Vinci's *Seven Principles*. The paper presents the educating AEC professionals is presented the AEC sector. The Theory of *Successful Intelligence* and its three forms of intelligence (Practical, Analytical, and Creative), are supported by lessons learned from the Renaissance, including the *Medici Effect* and da Vinci's *Seven Principles*.

## Results

Based on these theoretical pillars, a new approach to educating AEC professionals is presented with a proof-of-concept prototype that uses a game-like virtual reality (VR) visualisation interface supported by Mind Mapping is introduced as an exemplar.

## Conclusions

The developed interface in this study applies Game Theory to non-collocated design teams in accordance with Social Sciences Theory (social rules) and Behavioural Science Theory (decision making). It contributes by supporting new insights into AEC actor involvement, pedagogy, organisational behaviour, and the social constructs that support decision making.

# BACKGROUND

The Architecture-Engineering-Construction (AEC) sector is one of the largest industrial employers in many countries. In the European Union (EU) for example, it encompasses more than 2 million enterprises and approximately 12 million employees, representing 9.8% of the EU's Gross Domestic Product and employing over 7.1% of the workforce (NGRF 2010). This contribution and global competition makes the novelty of the AEC projects increasingly important. Therefore, AEC professionals need to be educated how to develop not only traditional, or routine projects, but also projects incorporating novel designs and construction processes. They need to be creative, and be able to develop unknown (or unproven) solutions which are feasible, surprising, and potentially patentable. Currently, AEC professionals are no longer being seen as leaders or innovators, more followers - using deductive problem solving rather than seeking opportunities, using their creativity and developing inventions. This resonates with thinking derived from innovation literature (Akintoye et al. 2012). As a result, designers and engineers in particular have seemingly lost their ability to innovate. This is partly attributable to 'inappropriate' education that has historically focused on production, rather than creativity. This is just the opposite of what happened in the 19th and early 20th Centuries, when designers and engineers were seen as the true 'drivers' of change. During this time, high-level education was aligned to incentives (e.g. the highest salary rates) which helped design and engineering schools attract the most talented students; and these graduates were capable of meeting all technological and socio-cultural challenges of the quickly expanding societies (Arciszewski 2006; Arciszewski and Harrison 2010a, 2010b; Arciszewski and Rebolj 2008). For instance, the construction of some monumental buildings during this period in history (e.g. Eiffel Tower, Villa Savoye, and The Bauhaus Building) created not only technological solutions, but also cultural revolutions-leading to a fundamental change in the way design and engineering was perceived.

This research posits that creativity has increasingly been underrepresented; and as such, needs to be revisited, especially in a rapidly evolving technological-driven world. For example, such challenges now include environmental and sustainability demands, increased levels of safety compliance, enhanced security issues, and whole life demands (energy, maintenance etc.). Whilst it could be argued that some of these challenges extend beyond the AEC domain *per se*, it is important to identify the key promoters and inhibitors of engineering creativity. In doing so, the profession as a whole will benefit from a new cogent way of embedding creativity into solutions; the result of which will not only benefit society, but also help inspire future AEC successors to follow this approach. Any changes, particularly those related to the ways that AEC students are educated, are extremely difficult, mostly because of the Vector of Psychological Inertia (G. Altshuller 1984) in action. This phenomenon refers to a natural tendency of individuals and communities to resist any changes, thereby delaying progress as much as possible. This is also influenced by the way in which the instructors were originally educated (mostly as highly sophisticated analysts) as this has a significant impact on the way they want to teach students. Cognisant of this, it is important to recognise the need to apply a complex systems approach to analyse the impact of this in order address the current situation.

This paper presents design and engineering leadership as three interrelated abilities: 1) to develop a vision, 2) to transform it into a strategy, and 3) to implement it. The key to leadership is the ability to develop feasible ideas or concepts (e.g., a new type of engineering system or construction process) using a set of abilities (traits) required to implement them [as opposed to using existing concepts to perform typical/routine work]. In particular, the development of a vision similar to conceptual design, to inventive design. In both cases a new idea, or a concept of an engineering system, needs to be developed. This is the area of activities in which creativity, or abductive generation of new ideas, takes place. This position is proffered, as historically, 'followers' have been seen to create stagnation, producing what has been called

"Vector of Psychological Inertia" (H. Altshuller 1984), or fixation (Youmans and Arciszewski 2014). This psychological phenomenon therefore tends to makes change and progress more difficult, and in some cases often even prevents it. The emphasis therefore is to consider the development of leaders (not followers), in order to minimise the negative impact of the Vector of Psychological Inertia.

Building upon the principles of the *Theory of Successful Intelligence (Sternberg1985,1996, 1997)*, this paper describes "Success" as a relative concept, which is defined by a given person in relation to the socio-cultural context and personal desires. This study therefore posits that there is a need to develop a new paradigm that recognises the importance of both analytical and creative works. Given this, this research defines analyst learners as the people who use rote learning and deduction, eventually induction, as opposed to creative people who use also abduction for reasoning. This approach extends learning capability beyond the learners' cognitive capability. Relying on the principles of Theory of Successful Intelligence (Sternberg 1985, 1996, 1997), Positive Psychology (Schueller 2012), and Appreciative Intelligence (Barrett and Fry 2008), this paper asserts that by using the 'right' methodologies and media, general principles of creative work could be translated into an explicit knowledge form and become part of a body of knowledge; hence, enabling the "Successful Departments" (Arciszewski 2009) to teach learners the *"Creative Intelligence"* and "Appreciative Intelligence". In this context, the potential of utilising advanced visualisation tools such as immersive game-like virtual reality interfaces is deemed vital - especially for augmenting analytical and parametric thinking capacity to intuitive idea generation (which could both be supported by these interfaces).

## Theory of Successful Intelligence

The *Theory of Successful Intelligence* (Sternberg 1985, 1996, 1997) is a major step toward understanding how individuals' abilities are interrelated with their life success. In the context of design and engineering education, this theory presents a new understanding of how education can be conceptualised, designed, and delivered. Through this theory, successfully intelligent people are defined as those being able to achieve their goals by leveraging their strengths, by compensating for their weaknesses, and those able to adapting to, shape, and select environments that will facilitate their success. This theory is underpinned by three fundamental pillars:

1.  Successful intelligence can be learned;
2.  Successful intelligence is a combination of three independently acquirable abilities, namely: practical intelligence, analytical intelligence, creative intelligence;
3.  Successful intelligence is dynamic; both the criteria of success and the abilities the individual employs (i.e. the relative combination of the three intelligences) to achieve success may change during one's life-time.

In accordance with this theory, practical intelligence is an ability to solve simple everyday problems, and this is achieved using readily available knowledge and heuristics. Abilities to open a door or to ride a bus are good examples of practical intelligence. Analytical intelligence is an ability to solve analytical problems, and that requires using deductive skills and utilising existing knowledge (for example, analysis of traffic flow, numerical optimization, or planning a typical construction process, etc.). Analytical intelligence is acquired through the combination of rote learning and learning deductive skills. Analytical intelligence alone is what traditional IQ tests measure. In addition, traditional engineering education emphasizes analytical intelligence almost entirely. However, the Theory of *Successful Intelligence* stipulates that a balance of the three intelligences is absolutely necessary for life success, including professional success.

In the AEC context, creative intelligence is the ability to solve inventive problems, which require abductive skills and obviously the use of existing knowledge. Solving such problems requires development of unknown solutions or ideas, e.g. development of a new type of a wind bracing system in a tall building or a new type of a tunnel. Creative intelligence is acquired through the combination of rote learning with learning of both deductive and abductive skills.

# The Medici Effect

Johansson (2004) proposed two interrelated concepts of the "Medici Effect" and of the "Intersection". These concepts identify mechanisms driving an environment facilitating and stimulating emergence of transdisciplinary knowledge, which is the foundation of creativity in engineering. Transdisciplinary knowledge is a body of integrated knowledge with roots in two or more domains, but knowledge which is domain-independent. For this reason, both concepts are important for engineering educators who should recreate them when educating creative engineers.

The Medici Effect (Johansson 2004) was a mechanism driving the emergence of the Renaissance intellectual foundation. It was named after the Medici family, which lived in Florence, Italy, in the 15th Century. The Medicis were one of the richest families in Europe, and sponsored many artists and scientists who were members of their court. Ultimately, members of the community began developing understanding of knowledge from outside of their domains. That led to new understanding of individual disciplines and to gradual emergence of the transdisciplinary knowledge. This knowledge became the intellectual foundation of the Renaissance.

The Intersection (Johansson 2004) is a product of the Medici effect. Johansson (2004) argued that Intersection is a time and place specific integration of knowledge with elements coming from various disciplines, cultures, and personalities. When a new concept is developed within a given discipline, it usually follows the existing line of evolution (Zlotin and Zusman 2006) and is considered directional. However, when an intersection occurs, a new idea represents a radical change, or beginning of a new line of evolution. Such an idea can be called "intersectional idea". Intersection can be described as knowledge integration with knowledge coming from two or more domains and resulting in transdisciplinary knowledge, valid in all contributing domains (Sage2000).

The Medici Effect should be used in AEC education to create an educational environment, called by Arciszewski (2009) "Successful Department". Such an environment should be not only supporting but also stimulate teaching and learning engineering creativity. This is a way to reconstruct an environment critical for the emergence of the Renaissance; and more importantly, for the creation of an environment necessary to educate creative designers and engineers.

# Da Vinci's Principles

Gelb (1998, 1999, 2004) introduced the term "Da Vinci Seven Principles" and proposed seven principles describing the core characteristics of Da Vinci's approach to science, design and engineering. These seven principles which are shown in an artist's vision in (Figure 1) are as follows:

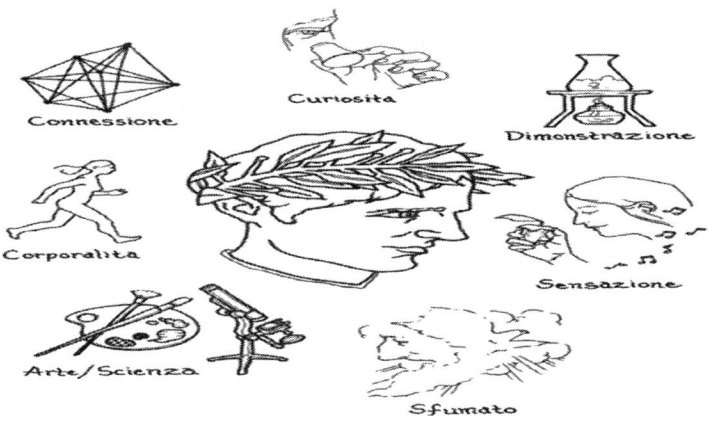

**Figure 1:** Da Vinci Seven Principles, Source: (Arciszewski 2009).

*Principle No. 1, "Curiosita,* means in Italian "curiosity." According to Gelb (2004), this da Vincian represents a curious and open attitude and a life-long learning accomplished by constantly asking questions about everything. Unfortunately, mostly analytical current educational materials gradually destroy *Curiosita* (Arciszewski 2014). In order to educate creative designers and engineers, it is necessary to not only maintain their *Curiosita,* but also expand it. The challenge here is to teach students the practical and analytical intelligence and at the same time expand their *Curiosita,* which is the key to the creative intelligence.

Principle No. 2, "Dimostrazione", means in Italian "demonstration". Gelb (2004) explained it as a unique attitude of experimentally verifying acquired knowledge. The Renaissance concept of apprenticeship is a good example of Demonstration in education. It was a combination of individual studies with extensive hands-on experience resulting in experiential learning. In this case, a master/teacher provides only guidelines and helps students how to learn on their own.

Principle No. 3, "Sensazione" means in Italian, "sensitivity to feelings". Gelb (1998) used this term to identify a complex attitude. It is development of all senses, practicing both the rational/intellectual and emotional approaches to life, and problems, integration of all abstract and physical inputs to create synaesthesia. It is a complex emotional state when an artist or a scholar is fully engaged in solving a problem, both intellectually and emotionally - using all his/her senses as using synaesthesia to acquire transdisciplinary knowledge or to create new ideas. Sensazione can be also interpreted as a practicing "whole-brain thinking" in which focus is on the emotional dimension of our cognition ultimately leading to a much more complete understanding of the world, of our environment, and of ourselves, including our consciousness and ability to transform it.

Principle No. 4, "*Sfumato*," means in Italian "turn to mist"; or in the case of colours, "soft," or "mellow." Gelb (1998) interpreted *Sfumato* as a willingness to accept and to understand the world in its infinite complexity. It also means keeping an open mind in the face of unknown and of uncertainty, a willingness to embrace contradictions, and paradoxes, and acquiring, accepting and using ambiguous knowledge. Sfumato is a surprisingly modern notion. In the case of knowledge discovery and inventive problem solving such processes are known today as lengthy and having subsequent periods of conscious and subconscious activities. To produce inventions, all kinds of input are obviously desired in order to activate and use the entire power of the human brain, both the analytical left hemisphere and the creative right side.

Principle No. 5, "*Arte/Scienza*" means that a Renaissance person should be a "Whole-Brain Thinker". People should develop an understanding of the world using two entirely different but complementary perspectives with roots in art and science, respectively. These two perspective should be balanced since both are necessary but neither sufficient. If AEC educators are interested in creating an education producing inventors, then the Principle "Arte/Scienza" is very important. It represents a significant departure from the traditional engineering education nearly entirely focused on the rational, "scientific" approaches and knowledge.

The Principle No. 6, "*Corporalita*" means "the state of being in physical or bodily form rather than spiritual form" in accordance to

MSN Encarta. It is incomplete, if not simply wrong description of the Da Vinci's Principle No. 6. The attitude of *Corporalita* is much more complex. It is described by Gelb (1998) as *"Means sana in corpora sano"* - sound mind in a sound body. Da Vinci argued that a human being should carefully maintain a balance between the intellectual and physical development in order to realize his/her full potential. It was also a reflection of the Renaissance belief that a genius must be physically superior with respect to ordinary people. *Corporalita* is particularly important for designers and engineers. They need to maintain a balance between body, mind, and spirit, but also to attain a relatively high level of physical fitness to survive long hour of climbing stairs and ladders on a construction site.

Principle No. 7, *"Connessione"* means in Italian "connection", however this principle in fact means recognition of interconnectedness of all things and phenomena in nature and life, the world is a single system with its all elements connected by direct and indirect feedbacks, the world is a complex and chaotic system, knowledge is a non-linear system. Only recently, in the second half of the 20th Century, the science of holistic understanding of the world, called "Cybernetics" emerged. It gradually led to the development of the "Systems Analysis" based on the principle of wholeness in its approach to analysis of all systems, built and natural, real and abstract, small and large. Therefore, *"Connessione"* may be interpreted as a systems view of the world.

# METHODS

## Successful Education

Successful Education (Arciszewski 2009) is a new paradigm in design and engineering education. This paradigm was inspired by the latest developments in the modern cognitive psychology, especially by the Theory of *Successful Intelligence* (Sternberg 1985, 1996, 1997). This paradigm has also been strongly influenced by a new understanding of historical and social mechanisms behind the emergence of the Renaissance, including the Medici Effect (Johansson 2004) and the Da Vinci Principles (Gelb 1998, 1999, 2004). (Arciszewski (2009)) argued that Principles are particularly important because they provide a synthesis of all attitudes practiced by Da Vinci and by the other great Renaissance engineers.

In this paradigm, the key concept is "Successful Designers and Engineers" and it describes the designers and engineers who have acquired as students not only the necessary and sufficient body of knowledge to practice engineering, but also learned *Successful Intelligence* including its all three components, i.e. practical, analytical, and creative intelligence. Such graduates are prepared to not only undertake any kind of routine work, but, if necessary, also prepared to become inventors and leaders, since in both cases the key to success is an ability to develop new ideas.

In Table 1, Successful Education is compared with a past design and engineering education paradigm, called "Master-Apprentice Paradigm", and the present one, called by us "Scientific Paradigm". The comparison is done from the perspective of the Theory of *Successful Intelligence* and of its three main components. In this context, only Successful Education is complete since only it addresses all three components of *Successful Intelligence* and consequently creates an opportunity to educate successful engineers.

**Table 1:** Comparison of teaching paradigms

| Teaching paradigm | Practical intelligence | Analytical intelligence | Creative intelligence |
|---|---|---|---|
| Master-apprentice | Yes | | Yes |
| Scientific | Yes | Yes | |
| Successful education | Yes | Yes | Yes |

Pour Rahimian et al.

Pour Rahimian et al. Visualization in Engineering 2014 2:4, doi:10.1186/2213-7459-2-4

Successful Education requires not only a new understanding of design and engineering education priorities and several new or modified courses, it also requires a complex environment, called "Successful Department", which will enable and stimulate the creation of successful engineers. A modern Medici Effect and the resulting intersection of ideas are crucial for the learning process. Therefore, they require a revolutionised environment (in terms of intellectual and technological structures) which is completely different that the current look of so many design and engineering departments. In essence, there are four major components of a Successful Department, namely courses, instructors, physical environment, and ambience (Arciszewski 2009). This is aligned with Salama's (2008) *"Integrating Knowledge in Design Education"* theory which argues that a responsive architectural design pedagogy giving credit to socio-cultural, and environmental needs can enable future architects to create livable environments.

Traditional, analytical courses are absolutely necessary for the future successful engineers, although they are grossly insufficient for them. They require additional courses on Inventive Design and Engineering, i.e. focused on the emerging science of inventive problem solving. For the best results, such courses could/should be offered to students through their entire period of studies. A single course for seniors (the present practice at George Mason University) is a step in the right direction, but it comes too late to impact learning in other courses and to transform students into successful engineers. A much better solution is a sequence of several courses, even if the total number of credit hours is the same.

Instructors are the key component of a Successful Department. A faculty in academic units are surprisingly similar in many aspects (ergo birds of feather flock together) despite all efforts to create diversity, which is often imposed only for political reasons. A successful Department requires, however, a true diversity, which may be described as "balanced intersection". This term is understood as a selection of instructors resulting in a department in which cultural backgrounds of instructors are strongly differentiated, they represent both applied and fundamental research, have experience in analytical and exploratory research and they represent various thinking styles.

Physical environment creates a framework for learning and also send a message about the nature of a given academic unit (Hou and Ji 2010). An ideal urban design for a Successful Department should be based on the concept of the agora, as an ideal form stimulating human interactions through complex socio-psychological mechanisms. Such an urban complex should have several buildings, arranged around the central square/agora. A building should be dedicated to teaching practical intelligence and designed with all kinds of testing laboratories and workshops. Another building should be dedicated to teaching analytical intelligence and it should have various computer laboratories. A third building, "Inventors Heaven", a must, should be dedicated to teaching creative intelligence with appropriately selected laboratories and workshops specifically designed for teams working on their inventive challenges. Finally, there should be an administrative building for faculty and classrooms.

A Successful Department would never be fully effective without a proper ambience. In this case, ambience is understood as a multi-sensory experience that positively affects students, faculty, and staff helping them to learn or teach in the best way to create successful engineers. Ambience obviously has an emotional dimension, which distinguishes it from a traditional department. Ambience is a reflection of people's perception of an environment surrounding them and can be carefully created in such a way as to contribute successful designers and engineers. Arciszewski (2009) discussed various components of ambience in a Successful Department, e.g. guiding principles and stories, colours, music, art, various activities, and even the proper lighting in the Successful Department.

Building upon the theoretical bases discussed in the theory of Successful Education (Arciszewski2009), this paper highlights the potentials of the advanced IT interfaces for leveraging all four components of such a Successful Department. The paper particularly suggests use of advanced game-like virtual workspaces in order to leverage education of successful designers and engineers for the AEC professions.

## Games and Virtual Reality in Construction Engineering Education

The nature and complexity of communication mechanisms within the Architecture, Engineering, and Construction (AEC) projects has changed significantly over the last ten years, especially the modus operandi and integration with core business operations. This has been reflected through the increased prevalence, use, and deployment of web-based project collaboration technologies and project extranets. Within the AEC sector, Information and Communications Technology (ICT) has revolutionised production and design (Cera et al. 2002), which has led to dramatic changes in terms of labour and skills

(Fruchter 1998). However, it is also important to acknowledge that the capabilities of such applications (and implementation thereof) in predicting the cost and performance of optimal design proposals (Petric et al. 2002) should enable design engineers to compare the quality of any one tentative solution against the quality of previous solutions. This was further reinforced by Goulding et al. (2007), regarding the ability to experiment and experience decisions in a 'cyber-safe' environment in order to mitigate or reduce risks prior to construction. It is therefore crucial for the AEC industry to employ cutting-edge ICT technologies to issues related to organisational management and decision making (Friedman 2005). Furthermore, whilst advocates note that these have helped to resolve some of the aforementioned challenges, Pour Rahimian et al. (2011) noted that project teams are still facing real and signification problems and challenges regarding heterogeneous systems faced by project teams using project extranets. In this essence, the problem here is that the industry is experiencing confusion as to how to manage project information in order to support decision-making processes. This is the point where Fruchter (2004) suggested the digital integration of the whole data creation, retrieval, and management system within building industry in order to prevent tacit knowledge loss and miscommunication among various parties from different disciplines. In this respect, recent innovation in Virtual Reality (VR) technologies and AEC decision- support toolkits have now matured, enabling tele-presence engagement to occur through integrated collaborative environments. Several opportunities are now available, including significantly improved immersive interactivity with haptic support that can enhance users' engagement and interaction.

Employing cutting edge ICT tools is also expected to leverage training systems within the AEC sector (Fruchter 1998) as the implementation of effective training could make impact on the whole industry by addressing and fulfilling the needs of the different stakeholders in the industry. In this respect, advanced ICT systems are expected to address the shortcomings of 'typical' learning models that often provide the trainees with only general instructions (Laird 2003) and issues associated with unaffordable costs of the 'traditional' on -the-job trainings (Clarke and Wall 1998). Therefore, new ICT advancements that incorporate innovative proactive experiential learning approaches which link theory with practical experience, using Virtual Reality interactive learning environments can be especially effective (Alshawi et al. 2007). This research builds upon the findings of previous studies in this area and links them to the principles of Successful Education (Arciszewski 2009), with specific emphasis on supporting the decision-making process at the construction stages. The study provides a novel approach of applying Game Theory to non-collocated design teams using Game-Like VR environments blended to Social Sciences Theory (social rules) and Behavioural Science Theory (Decision Science/Communication Science). In essence, the aim of this study is to advocate the advantages of applying flexible, interactive, safe learning environment for practicing new working conditions with respect to offsite production (OSP) in general, and Open Building Manufacturing (OBM) in particular; without the 'do-or-die' consequences often faced on real construction projects (Goulding and Rahimian 2012).

As the underpinning technology, VR has been defined as a 3D computer-generated alternative environment to be immersed in, for navigating around and interacting with (Briggs 1996), or as a component of communication taking place in a 'synthetic' space, which embeds human as its integral part (Regenbrecht and Donath 1997). The definitions of VR systems usually includes a computer capable of real-time animation, controlled by a set of wired gloves and a position tracker, and using a head-mounted stereoscopic display as visual output. For instance, Regenbrecht and Donath (1997) defined the tangible components of VR as a congruent set of hardware and software, with actors within a three-dimensional or multi-dimensional input/output space, where actors can interact with other autonomous objects, in real time. VR has also been defined as a simulated world, which comprises of some computer-generated images conceived via head mounted eye goggles and wired clothing – thereby enabling the end users to interact in a realistic three-dimensional situation (Yoh 2001).

Over the last 30 years, ICT systems have matured and enabled construction organisations to fundamentally restructure and enhance their core business functions. A.Z. Sampaio and Henriques (2008) asserted that the main objective of using ICT in construction field is supporting management of digital data, namely to convert, store, protect, process, transmit, and securely retrieve datasets. They acknowledge the commencement of VR techniques as an important stepping stone for data integration in construction design and management as they are capable of holding and presenting the whole information about buildings (e.g. size, material, spatial relationships, mechanical and electrical utilities, and etc.) through a single output. Similarly, Zheng et al. (2006) proposed the use of VR to reduce time and costs in product development and to enhance quality and flexibility for providing continuous computer support during development lifecycle.

Early studies that incorporated VR into the design profession used it as an advanced visualisation medium. Since as early as 1990, VR has been widely used in the AEC industry as it forms a natural medium for building design by providing 3D models, which can be manipulated in real-time and used collaboratively to explore different stages of the construction process (Whyte et al. 1998). It has also been used as a design application to provide collaborative visualisation for improving construction processes (Bouchlaghem et al. 2005). However, expectations of VR have changed during the current decade. According to A.Z. Sampaio and Henriques (2008), it is increasingly important to incorporate VR 3D visualisation and decision support systems with interactive interfaces in order to perform real-time interactive visual exploration tasks. This thinking supports the position that a collaborative virtual environment is a 3D immersive space in which 3D models are linked to databases, which carry characteristics. This premise has also been followed through other lines of thought, especially in construction planning and management by relating 3D models to time parameters in order to design 4D models (Fischer and Kunz 2004), which are controlled through an interactive and multi-access database. In similar studies, 4D VR models have been used to improve many aspects and phases of construction projects by: 1) developing and implementing applications for providing better communication among partners (Leinonen et al.2003), 2) supporting design creativity (Rahimian and Ibrahim 2011), 3) introducing the construction plan to stakeholders (Khanzade et al. 2007), and, 4) following the construction progress (Fischer 2000).

With regards to education, Wellings and Levine (2010) posited that there was a need to redesign the current text-based lessons into collaborative and multidisciplinary problem-based materials, expressly to take on board real world problems and solutions. They argued that this was not possible unless immersive and interactive games were employed for improving trainees' engagement. Similarly, Thai et al. (2009) asserted that pedagogical digital games offered an intact opportunity to enhance engagement of trainees and revolutionise teaching and learning. ACS (2009) summarised the benefits of the emerging educational interactive immersive game environments: 1) annotated objects could provide deeper level of knowledge on demand, 2) incorporating additional dimensions of subjects (nD), 3) supporting distance team collaboration, 4) leveraging equal opportunities by providing distance learning opportunities and, 5) simulated learning by modelling a process or interaction that closely imitates the real world in terms of outcomes.

VR applications and game engines are now increasingly being used in the teaching and learning AEC. According to Zudilova-Seinstra et al. (2009), VR as a teaching tool can contribute to the trainees' professional future by developing some learning activities beyond what is available in the conventional training systems. With respect to educational issues in the AEC industry, A. Z. Sampaio et al. (2010) argued that the interaction with 3D geometric models can lead to active learner thoughts which seldom appear in conventional pedagogical conditions. Moreover, Juárez-Ramírez et al. (2009) asserted that when augmented to 3D modelling, VR could lead to better communication in the process of AEC training. However, VR training environments have arguably not yet fully reached the potential of reducing training time, providing a greater transfer of expert knowledge; or supporting decision making. This was primarily down to the ways in which this technology was augmented. It is therefore argued that educational training tools need to 'engage' learners by putting them in the role of decision makers and 'pushing' them through challenges; hence, enabling different ways of learning and thinking through frequent interaction and feedback, and connections to the real world context (Goulding et al. 2007). Furthermore, it is postulated that paring instructional content with game features, could engage users more fully, hence, help to achieve the desired instructional goals. In this respect, this study applied an input-process-output model (Garris et al. 2002) of instructional games and learning to design an instructional program which incorporated certain features or characteristics from gaming technology; which trigger a cycle that includes user judgment or reactions, such as enjoyment or interest, user behaviour such as greater persistence or time on task, and full learner feedback (Figure 2).

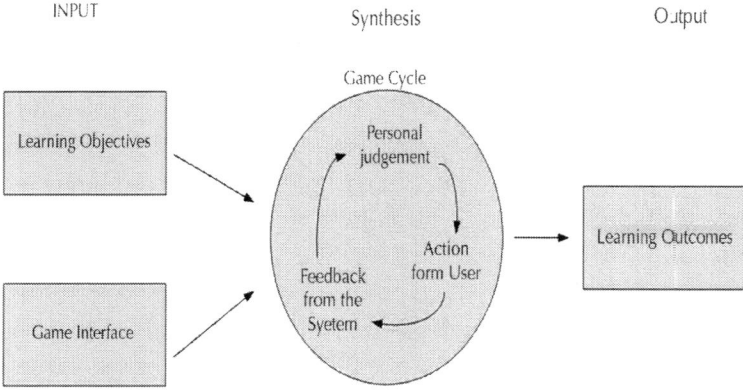

**Figure 2:** Educational game model input-synthesis-outcome (Garris et al.2002).

# RESULTS AND DISCUSSION

This section presents the developed Game-Like Virtual Reality Construction-Site Simulator (GVRSS) in this study. The aim of the developed GVRSS was to embrace 'real life' issues facing offsite construction projects in order to appeal to professionals by engaging and challenging them to find 'real life' solutions to problems often encountered on site. Given this, a real construction project was used to govern the authenticity of the learning environment. In this context, the prototype learning simulator was designed specifically to allow 'things to go wrong', and hence, allow 'learning through experimentation' or 'learning by doing'. In this respect, although the 'scenes' within the simulator take place on a construction site, the target audience was focussed primarily on construction professionals e.g. project managers, construction managers, architects, designers, commercials, suppliers, manufacturers etc. Thus, the construction site was used as the main domain through which all the unforeseen issues and problems (caused through upstream decisions, faulty work etc.) could be enacted. The key learning impact areas were to acknowledge the importance, significance and real implications of time, cost, resources etc. Learning was planned and reinforced through a debriefing session, where learners were able to demonstrate additional understanding, particularly with respect to mitigating such issues in future construction projects. In this context, learning occurred through the following:

- Learner autonomy - to make all decisions;
- Interactivity - environment provides feedback on the decisions taken, and their implications on the overall project (cost, time, resources, health and safety, etc.);
- Reflection - users are able to defend decisions on the feedback provided, and have the ability to identify means to avoid/mitigate potential problems in the future.

In essence, the main concept of the simulator was based on its ability to run scenarios through a VR environment to address predefined training objectives. In this respect, learning was designed to be driven by problems encountered in this environment, supported by a report critique on learners' choices, rationale, and defence thereof. In accordance to these objectives, the GVRSS was designed and developed as an educational web-based simulation tool comprising of both non-immersive and immersive pages for providing construction managers (and other disciplines) the opportunity to experience challenges of real-life AEC projects through simulated scenarios. In order to minimise interruption on the learners' reasoning process, the Graphical User Interface (GUI) was designed to be as simple and straightforward as possible with respect to data input. Thereby, the interface was designed as to be accessible through any standard web browser to provide users with login account details and other criteria, e.g. selection of available construction sites, projects, contractors, equipment, scenarios etc. All choices made by 'players' as well as their registration data was automatically recorded in a MySQL database, which was also accessible through the immersive application for project simulation. After completing the initial decision-making process through the interactive ASP. Net Web Forms, learners are able to commence the training session, starting with a 'walkthrough' to experience and appreciate the complexity of the project. At this stage, the application provides users with a summary of the project and contract, and runs the simulation of the project within an immersive and interactive environment developed in Quest3D™ VR programming Application Programming Interface (API). Within the simulated Quest3D environment, the users are able to experience the outcomes of all decisions made. They are also challenged by unexpected events designed according to the selected scenario, and are required to make decisions for dealing with these issues. The monitoring and communication tools are embedded in different parts of the main interface as well as the facilitated standard embedded virtual PDA or smart phone-type interface, which appears when required. The simulator ultimately records and tracks the users in the database and navigates to the conclusion page to reveal all scores of the user (together with the logic behind the marking procedure). Figure 3 illustrates a selection of the various functions available to the user of the simulator to fully interact with and retrieve information from the simulator during the VR simulation session. Further inclusion of the whole tree is considered for the exploitation phase.

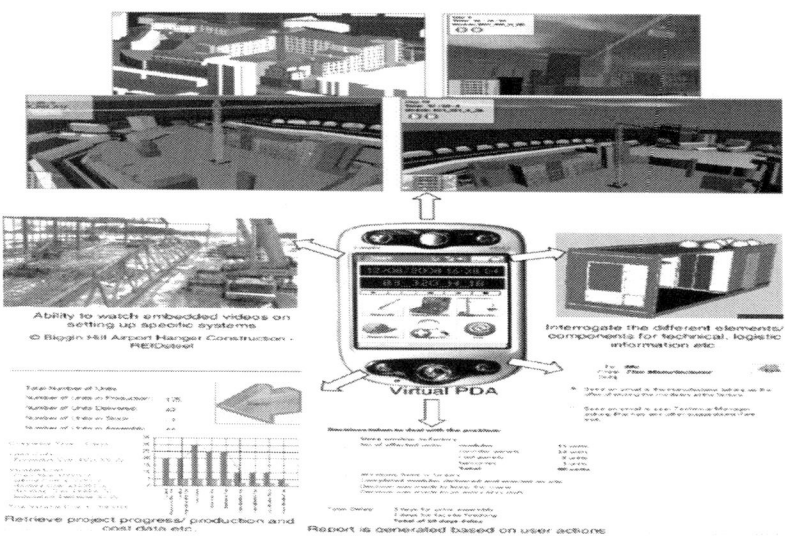

**Figure 3:** The VR simulation sessions.

# CONCLUSIONS

Construction projects are increasingly becoming more complex, often engaging new business processes and technological solutions to meet ever-increasing demands. These business demands are complex and multifarious; often requiring the conjoining of high level skill sets to deliver the solutions needed. These skill sets are currently underrepresented, and seldom engage the collective ethos needed to envelop creative thinking, through such approaches as *Successful Intelligence* in order to create new innovative solutions. It is therefore paramount that the industry as a whole engages the right type (and level) of skill sets and competence needed to meet these project requirements and business imperatives. Acknowledging this, it is also important the causal drivers and influences associated with creativity and successful decision-making in global AEC teams are fully understood and supported. This however, requires a radical review in the way educational programmes and systems are designed and delivered. For example, with respect to leveraging creativity and delivering innovation, this study reflected on the Renaissance period and the creativity-oriented learning/teaching paradigm called *"Master-Apprentice Paradigm"*, as opposed to the current analysis focused *"Science Paradigm"*. The Medici Effect and the related phenomenon of intersection and seven *Da Vinci Principles* have been acknowledged as being able to revolutionise modern design and engineering education. This study then introduced the theory of *Successful Intelligence* and its three components as an underpinning platform for educating a new generation of designers and engineers.

The *"Successful Education"* paradigm (Arciszewski 2009) was presented as a new approach for educating AEC professionals was presented, including the concept of a new educational environment; the need for a new combination of courses that focus on teaching the three kinds of *Successful Intelligence* (in the context of AEC sector); including guidelines of how to properly select instructors that are capable of implementing such approach. A proof-of-concept prototype that uses a game-like virtual reality (VR) visualisation interface supported by Mind Mapping was presented as an exemplar, to demonstrate how the proposed approach could be implemented. The developed simulator offers a risk free environment where learners can evaluate how decisions they make affect their business. This includes (but is not limited to) analysing issues occurring on the construction site, such as: design concerns, process conflicts, logistics challenges, and supply chain issues etc.

This paper proffers that enhanced engagement through an immersive project environment could lead to a better understanding of the real-life AEC problems. This can be achieved by placing learners in a cyber-safe environment; specifically to leverage learners' cognitive processes to real-world issues. This study supports a novel approach of applying Game Theory to non-collocated design teams using Game-Like VR environments blended to Social Sciences Theory (social rules) and Behavioural Science Theory (decision science/communication science). This can address the need to evaluate actor involvement in order to reveal new insight into AEC organisational behaviour and the social constructs that often affect decision making. In this paper, advanced VR training and simulation tools were proffered through an exemplar in order to highlight the possibilities available, especially as this forms a conduit for aligning pivotal drivers to achieve specific learning outcomes. Future research in this area is likely to embrace the importance of pedagogy (learner styles/traits), as this has been openly acknowledged as being particularly efficacious and instrumental for delivering training material to specific learner-types.

# AUTHORS' CONTRIBUTIONS

TA developed the theory of Successful Education. JS Goulding developed the case study of Game-Like Virtual Reality Construction-Site Simulator. FPR carried out further programming and coding for extending the Game-Like Virtual Reality Construction-Site Simulator in order to enable non-collocated AEC collaboration through this interface. All three authors worked on linking the principles of theory of Successful Education to the potentials of emerging VR interfaces. All authors read and approved the final manuscript.

# REFERENCES

1.   ACS (2009). 3D Learning and Virtual Worlds. An ACS: Expertise in ActionTM White Paper. http://www.trainingindustry.com/media/2043910/acs%203d%20worlds%20and%20virtual%20learning_whitepaper%20april%202009.pdf, Access Date, 22/02/2014

2.   Akintoye A, Goulding JS, Zawdie G (2012). Construction Innovation and Process Improvement London: Wiley-Blackwell.

3.   Alshawi, M, Goulding, JS, Nadim, W In Kazi AS, Hannus M, Boudjabeur S, Malon A (Eds.) (2007). Training and Education

for Open Building Manufacturing: Closing the Skills Gap. *Open Building Manufacturing: Core Concepts and Industrial Requirements* Helsinki, Finland: ManuBuild in collaboration with VTT - Technical Research Centre of Finland.

4. Altshuller, G (1984). Creativity as an Exact Science. New York: Gordon and Breach, Science Publishers, Inc...

5. Altshuller, H (1984). Creativity as an Exact Science. New York: Gordon and Breach, Science Publishers, Inc...

6. Arciszewski, T (2006). Civil Engineering Crisis. *ASCE Journal of Leadership and Management in Engineering, 6*(1), 26–30.

7. Arciszewski, T (2009). Successful Education. How to Educate Creative Engineers. Fairfax, VA: Successful Education LLC...

8. Arciszewski, T (2014). Future of engineering education. *Proceedings of the ICE - Management, Procurement and Law, 167*(1), 46–59. Publisher Full Text

9. Arciszewski, T, & Harrison, C (2010). Successful Civil Engineering Education. *ASCE Journal of Professional Issues in Engineering Education and Practice, 136*(1), 1–8.

10. Arciszewski, T, & Harrison, C (2010b). Successful Education: The Key to Engineering Creativity (Paper presented at The International Conference on Computing in Civil and Building Engineering). Nottingham, UK: Nottingham University Press.

11. Arciszewski, T, & Rebolj, D (2008). Civil Engineering Education: Coming Challenges. *International Journal of Design Science and Technology, 14*(1), 53–61.

12. Barrett, FJ, & Fry, RE (2008). Appreciative Inquiry: A Positive Approach to Building Cooperative Capacity. Chagrin Falls, OH: Taos Institute.

13. Bouchlaghem, D, Shang, H, Whyte, J, Ganah, A (2005). Visualisation in architecture, engineering and construction (AEC). *Automation in Construction, 14*(3), 287–295.

14. Briggs, JC (1996). The Promise of Virtual Reality. *The Futurist, 30*, 30–31.

15. Cera, CD, Reagali, WC, Braude, I, Shapirstein, Y, Foster, C (2002). a Collaborative 3D Environment for Authoring Design Semantics. *Graphics in Advanced Computer-Aided Design, 22*(3), 43–55.

16. Clarke, L, & Wall, C (1998). UK construction skills in the context of European developments.*Construction Management and Economics*, *16*(5), 553–567.

17. Fischer, M (2000). 4D CAD-3D models incorporated with time schedule, CIFE Centre for Integrated Facility Engineering in Finland, VTT-TEKES, CIFE Technical Report, Helsinki. Finland: University of Helsinki Press.

18. Fischer, M, & Kunz, J (2004). The Scope and Role of Information Technology in Construction. CIFE Technical Report. (p. 19). San Francisco: Center for Integrated Facility Engineering, Stanford University.

19. Friedman, TL (2005). The World is flat: A Brief History of the 21st Century. New York: Farrar, Straus and Giroux.

20. Fruchter, R (1998). nternet-based Web Mediated Collaborative Design and Learning Environment, in Artificial Intelligence in Structural Engineering. Lecture Notes in Artificial Intelligence. (pp. 133–145). Berlin: Heidelberg: Springer-Verlag.

21. Fruchter, R (2004). Degrees of Engagement in Interactive Workspaces. *International Journal of AI & Society*, *19*(1), 8–21. doi:10.1007/s00146-004-0298-x

22. Garris, R, Ahlers, R, Driskell, JE (2002). Games, Motivation, and Learning: A research and Practice Model. *Simulation Gaming*, *33*(4), 441–467. Publisher Full Text

23. Gelb, MJ (1998). How to Think like Leonardo da Vinci. New York: Random House.

24. Gelb, MJ (1999). How to Think like Leonardo da Vinci, Workbook. New York: Random House.

25. Gelb, MJ (2004). Da Vinci Decoded: Discovering the Spiritual Secrets of Leonardo's Seven Principles. New York: Bantam Dell.

26. Goulding, JS, & Rahimian, FP In Akintoye A, Goulding JS, Zawdie G (Eds.) (2012). Industry Preparedness: Advanced Learning Paradigms for Exploitation. *Construction Innovation and Process Improvement* (pp. 409–433). Oxford, UK: Wiley-Blackwell.

27. Goulding, JS, Sexton, M, Zhang, X, Kagioglou, M, Aouad, GF, Barrett, P (2007). Technology adoption: breaking down barriers using a virtual reality design support tool for hybrid concrete. *Construction Management and Economics*, *25*(12), 1239–1250.

28. Hou, Y, & Ji, L (2010). Stimulating Design Creativity by Public Places in Academic Buildings. *Journal Structure and Envirinment*, 3(2), 5–13.

29. Johansson, F (2004). The Medici Effect. Boston, MA: Harvard Business School Press.

30. Juárez-Ramírez, R, Sandoval, G, Cabrera Gonzállez, C, Inzunza-Soberanes, S (2009). Educational strategy based on IT and the collaboration between academy and industry for software engineering education and training. Lisbon, Portugal, Badajoz, Spain: FORMATEX.

31. Khanzade, A, Fisher, M, Reed, D (2007). Challenges and benefits of implementing virtual design and construction technologies for coordination of mechanical, electrical, and plumbing systems on large healthcare project. Maribor, Slovenia: University of Maribor Press.

32. Laird, D (2003). New Perspectives in Organisational Learning, Performance, and Change: approaches to training and development. USA: Preseus Books Group.

33. Leinonen, J, Kähkönen, K, Retik, AR, Flood, RA, William, I, O'Brien, J (2003). New construction management practice based on the virtual reality technology. In R. R. A. Issa, I. Flood, & W. O'Brien (Eds.), 4D CAD and Visualization in Construction: Developments and Applications (pp. 75-100). Tokyo: AA Balkema Publishers.

34. NGRF (2010). N. G. R. F. http://www.guidance-research.org/future-trends/construction/info. Accessed 14th Aug 2010

35. Petric, J, Maver, T, Conti, G, Ucelli, G (2002). Virtual reality in the service of user participation in architecture. Aarhus Denmark: Aarhus School of Architecture.

36. Pour Rahimian, F, Ibrahim, R, Wirza, R, Abdullah, MTB, Jaafar, MSBH (2011). Mediating Cognitive Transformation with VR 3d Sketching During Conceptual Architectural Design Process. *Archnet-IJAR, International Journal of Architectural Research*, 5(1), 99–113.

37. Rahimian, FP, & Ibrahim, R (2011). Impacts of VR 3D sketching on novice designers' spatial cognition in collaborative conceptual architectural design. *Design Studies*, 32(3), 255–291.

38.    Regenbrecht, H, & Donath, D In Bertol D (Ed.) (1997). Architectural Education and Virtual Reality Aided Design (VRAD). *Designing Digital Space - An Architect's Guide to Virtual Reality* (pp. 155–176). New York: John Wiley & Sons.

39.    Sage, AP In Somerville MA, Rapport D (Eds.) (2000). Transdisciplinarity Perspectives in Systems Engineering and Management. *Transdisciplinarity: Recreating Integrated Knowledge* (pp. 158–169). Oxford: EOLSS Publishers Ltd...

40.    Salama, AM (2008). A Theory for Integrating Knowledge in Architectural Design Education. *Archnet-IJAR, International Journal of Architectural Research, 2*(1), 100–128.

41.    Sampaio, AZ, & Henriques, PG (2008). Visual simulation of previous termcivil engineeringnext term activities: didactic virtual previous termmodels. Plzen, Czech Republic: Visualization and Computer Vision.

42.    Sampaio, AZ, Ferreira, MM, Rosário, DP, Martins, OP (2010). 3D and VR models in Civil Engineering education: Construction, rehabilitation and maintenance. *Automation in Construction, 19*(7), 819–828.

43.    Schueller, SM In Ramachandran VS (Ed.) (2012). Positive Psychology. *Encyclopedia of Human Behavior* (pp. 140–147). San Diego: Academic Press.

44.    Sternberg, RJ (1985). Beyond IQ: A Triarchic Theory of Intelligence. Cambridge: Cambridge University Press.

45.    Sternberg, RJ (1996). Successful Intelligence. New York: Simon & Shuster.

46.    Sternberg, RJ In Coleangelo NN, Davis GA (Eds.) (1997). A Triarchic View of Giftedness: Theory and Practice. *Handbook of Gifted Education* (pp. 43–53). Boston, MA: Allyn and Bacon.

47.    Thai, AM, Lowenstein, D, Ching, D, Rejeski, D (2009). Game Changer: Investing in digital play to advance children's learning and health. The Joan Ganz Cooney Center. http://www.joanganzcooneycenter.org/wp-content/uploads/2010/03/game_changer_final_1_.pdf, Access Date: 06/06/2014

48.    Wellings, J, & Levine, MH (2010). The Digital Promise: Transforming Learning with Innovative Uses of Technology. A white paper on literacy and learning in a new media age, Joan

Ganz Cooney Center at Sesame Workshop. http://dmlcentral. net/sites/dmlcentral/files/resource_files/Apple.pdf, Access Date: 06/06/2014

49.  Whyte, J, Bouchlaghem, N, Thorpe, A (1998). The promise and problems of implementing virtual reality in construction practice. The Life-cycle of Construction IT Innovations: Technology Transfer From Research To practice (CIB W78), Stockholm, 3–5 June, 1998.

50.  Yoh, M (2001). The Reality of Virtual Reality. Seventh International Conference on Virtual Systems and Multimedia (VSMM'01), Organized by Center for Design Visualization. Berkley , USA: University of California Berkley. IEEE. doi:0-7695-1402-2/01

51.  Youmans, R, & Arciszewski, T (2014). Design Fixation: Classifications and Modern Methods of Prevention. Artificial Intelligence for Engineering Design, Analysis and Manufacturing. in print

52.  Zheng, X, Sun, G, Wang, SW (2006). An Approach of Virtual Prototyping Modeling in Collaborative Product Design.

53.  Zlotin, B, & Zusman, A (2006). Directed Evolution: Philosophy, Theory, and Practice. Southfield, MI: Ideation International.

54.  Zudilova-Seinstra, E, Adriaansen, T, van Liere, R (2009). Trends in Interactive Visualization: State-of-the-Art Survey (Advanced Information and Knowledge Processing). London: Springer-Verlag.

# The Effect of Rapid Social Changes during Post-Communist Transition on Perceptions of the Human - wolf Relationships in Macedonia and Kyrgyzstan

Nicolas Lescureux and D John C Linnell

Norwegian Institute for Nature Research, Sluppen, P.O. Box 5685, Trondheim, 7485, Norway

## ABSTRACT

The relationship between humans and wolves is often associated with conflicts strongly linked with livestock breeding activities. However, as conflicts are often more intense than expected considering the magnitude of their economic impact, some authors have suggested that these conflicts are disconnected from reality and are mainly due to persistence of negative perceptions from previous times. To the contrary, we suggest that local people's perceptions are often linked to wolf behaviour through direct observations and interactions. We conducted ethnological investigations on human-wolf relationships in countries belonging to former USSR (Kyrgyzstan) and former Yugoslavia (Republic of Macedonia), subjected to rapid social changes impacting both livestock husbandry and hunting practices. Our studies revealed that changes in hunting and husbandry practices have led to modifications in the socio-environmental context and to the nature of wolf-human interactions. These changes have resulted in an increased vulnerability of local people to wolf damage and a concomitant reduced acceptance for wolves. All these changes contribute to changes in the perception of the wolf and to an increase in the perception of conflicts, even in countries where humans and wolves have continuously coexisted. Our study shows the dynamic nature of human-wolf relationships, the necessity to understand the broader socio-economical context in human-wildlife conflicts, and the challenge pastoralists are facing in a changing world.

# BACKGROUND

The carnivorous diet and need for large living areas of large carnivores has led to an age-old competition with humans for space and food, thereby generating a range of economic and social conflicts (Treves and Karanth 2003). The most common dimension of conflict between humans and large carnivores – especially wolves – remains livestock depredation, and this conflict has been responsible for motivating the past reduction in the number and distribution of large carnivores on a worldwide level (Mech 1995, Breitenmoser 1998, Kaczensky 1999). Among large carnivores, wolves appear as one of the most conflictful species wherever they occur and overlap with herding activities, especially with sheep breeding. Nowadays, in many countries, either rural land abandonment or conservation legislation or both of them are leading to the recovery of large carnivores in multiple-use landscapes (Linnell et al. 2001, Falcucci et al. 2007, Linnell et al. 2009) and accordingly many conflicts are currently increasing in several countries (Mech 1970, Fritts et al. 1997, Bangs et al. 1998, Breitenmoser 1998, Treves and Karanth 2003).

Economic impacts appear to be generally low at national levels. A review of large carnivore impact on livestock in Central Europe showed that in most areas less than 1% of livestock is taken by large carnivores (Kaczensky 1999). Therefore, considering that present day economic impacts and risks to human safety (Moore 1994, Røskaft et al. 2003, Røskaft et al. 2007) are not sufficient to explain the intensity of the negative perceptions and social conflicts currently surrounding large carnivores, some authors have proposed that conflicts with large carnivores reflect the long-term persistence of negative perceptions from earlier times (Clark et al. 1996a, 1996b, Kellert et al. 1996, Lohr et al. 1996, Fritts et al. 2003).

However, it is also recognized that economic impacts can be very intense on a local level (Kaczensky 1999) and it appears that in some countries they are far more important (e.g. in Romania, cf. Mertens and Promberger 2001), and this may also be influenced by the presence or absence of compensation systems. In addition, several studies show that risks to human safety, even if close to zero in the modern occidental world, used to exist in Europe in the past and still exist in some developing countries (Rajpurohit 1999, Comincini et al. 2002, Linnell et al. 2003, Løe and Røskaft 2004, Moriceau 2007). Therefore, even if it is possible that some negative perceptions are inherited from earlier time, it also appears that they could find their root in actual human – wolf relationships and their real negative aspects. In addition, observing that tolerance to large carnivores is lower where the tradition of living together with them has been lost, Boitani suggested that prolonged sympatry can lead to a form of coexistence where compromises are made by both species and conflicts are not perceived as being so intense (Boitani 1992, 1995). This more dynamic and interactive vision is partly supported by ecological studies revealing the impact human activities can have on wolf behaviour and ecology (Vilà et al. 1995, Ciucci et al. 1997, Theuerkauf 2003, Theuerkauf et al. 2003a, 2003b) even though this impact has to be assessed through more generalized and standardized studies (Theuerkauf 2009). Concerning the impact wolves can have on humans, it has been addressed through studies about wolves in history and mythology (see e.g. Lopez 1978, Coleman 2004, Walker 2005), the study of symbolism in our modern perceptions of wolves (Bobbé 1993b, Bobbé 1993a, Bobbé 2003) and sociological analysis of wolf impacts on human society (Brox 2000, Mauz 2005, Skogen et al. 2008, Doré 2010). All these studies clearly show the impact wolves can have on humans at different levels but they are rarely connected with studies on the human impact on wolf behaviour and ecology, apart from some historical studies using historical data on both humans and on wolves (Moriceau 2007,Alleau 2010). Therefore they rarely shed light on the processes through which the human – wolf relationships are constructed.

The relatively recent integration of animals into social sciences through the recognition of their status as actors and active agents able to influence human social life (Descola and Pálsson 1996, Ingold 1996,Latour 1996, Ingold 2000, Haraway 2003, Brunois 2005a, 2005b, Descola 2005) has permitted the development of new interdisciplinary approaches to human – large carnivore relationships spanning both scientific and local knowledge (Lescureux 2006). These new approaches reveal the influence large carnivore behaviour can have on human perceptions and activities, and also provide insights into the interactive and dynamic character of human – wolf relationships (Lescureux 2006, Lescureux and Linnell 2010).

If the wolf – human relationship is so dynamic, social, economic and political transitions in human society should also result in changes to the human – wolf relationship. The rapid social changes like the ones which have occurred in most eastern European and central Asian countries at the beginning of the 1990's provide an opportunity to assess this dynamic and interactive nature of human – wolf relationships. Indeed, our hypothesis is that these changes potentially affected the human – wolf relationships in two manners. Firstly, these social changes have a direct impact on wolf ecology and behaviour through the changes in human hunting (of both wolves and their prey) and livestock husbandry practices. Secondly, rapid social changes create a new ecological and socio-economical context in which the human perception of wolves and their place in this context will be affected, all the more as the ecology and behaviour of this animal change. In turn, the changes in human perceptions of wolves will influence wolf hunting and livestock husbandry practices. Given the complexity of these processes, the comparison between two different countries which both underwent rapid social changes but through different processes allows us to explore how two different contexts have influenced the way the human – wolf relationship changed along with the rapid social changes. In addition, this comparison offers the possibility to better understand the origins of human – wolf conflicts in different contexts and to disentangle the different components (material, symbolic, relational) of these conflicts, which is a necessary step to be able to address them adequately.

# STUDY AREAS

We conducted research in two countries where rapid social changes occurred after the fall of the USSR and the fall of Yugoslavia. These were respectively the Republic of Kyrgyzstan where ethnographic surveys were made between 2003 and 2006, and the Republic of Macedonia where surveys were made between 2007 and 2010.

Kyrgyzstan and Macedonia are both mountainous countries, but there are some differences between these countries. Kyrgyzstan is larger than Macedonia and has a lower population density, which can be explained by the topographic conditions and overall low productivity. Indeed, more than 70% of the country is above 2000 m high and the highest peak reaches to more than 7000 m. The two countries are well suited for sheep husbandry, as permanent pastures constitute 47% of Kyrgyzstan's total land area and 25% of Macedonia's total land area. The landscape of Kyrgyzstan is an open one, as forest only covers 5% of the country. This open landscape favours mutual observations between humans and wolves. This is not the case in Macedonia where 37% of the total land area is covered by forest. The wolf is a hunted species in both countries, without limits on the hunting season or quota. Rough estimations give a population of 4000 to 6000 wolves in Kyrgyzstan and 800 to 1000 wolves in Macedonia (Boitani 2003, Salvatori and Linnell 2005).

In Kyrgyzstan, the study was conducted in two different areas (see Figure 1). The first one consisted of several villages around At-Bashy in Naryn region. These villages are at an elevation of around 2,500 m.a.s.l, and the main activity is livestock breeding, with a few cultivated areas, used mainly for onions and potatoes. The second one was focused on villages around Bokonbaevo in the Issyk-Kul region. These villages are situated along the Issyk-Kul Lake, at an altitude of around 1,600 m.a.s.l. and enjoy a milder climate allowing agricultural activities other than livestock breeding. Some interviews were also conducted in high pastures away from villages. As for the rest of the country, these areas are only sparsely forested. They mainly consist of agricultural areas with open fields and meadows in the valley bottoms with some shrubs, mainly along rivers. The landscape quickly shifts to alpine meadows with increasing altitude. Despite the departure of numerous young people to the capital, the study villages had not become depopulated, and had even experienced a slightly positive growth due to a high birth-rate. Wolves have been continuously present in the country for centuries and are subject to bounty hunting.

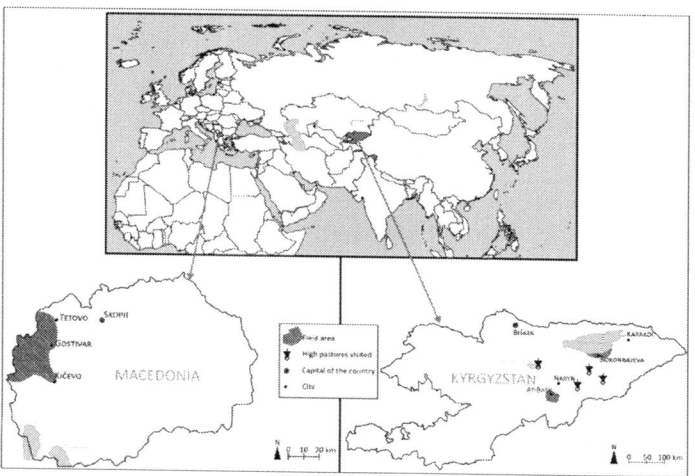

**Figure 1:** Republics of Macedonia and Kyrgyzstan in Eurasia, and Republics of Macedonia and Kyrgyzstan, showing field areas.

In Macedonia, the study was conducted in the Polog and Yugozapaden regions (municipalities of Tetovo, Gostivar and Mavrovo-Rostushe, see Figure 1). The area is predominantly rural, consisting of small towns and agricultural areas in the valley bottoms, with forested slopes, and alpine meadows at higher altitudes. The forests are widely used for hunting and forestry, and the alpine meadows are used by both transhumant and resident shepherds from May to October. Villages are scattered throughout the landscape, and have suffered dramatic and long term declines of the human population during recent decades, accelerating in recent years. Bears, wolves, and lynx have been continuously present during recent centuries in these regions of western Macedonia's mountains (Servheen et al.1998, Salvatori and Linnell 2005, Ivanov et al.2008). Both ethnic Albanian and ethnic Macedonians were interviewed. The population of wolves is still large and well connected with those in neighbouring countries.

# METHODS

This study is based on results from ten months of ethnological survey conducted between 2003 and 2007 in Kyrgyzstan and six months of ethnological survey in Macedonia between 2007 and 2011. In both countries we focused on two groups that appeared to be the most knowledgeable about, and the most involved in conflicts, with wolves: livestock breeders and hunters. In Kyrgyzstan our survey was mainly focused on a few villages since livestock breeders are still numerous, allowing us to make some participant observations, living with a family in the village for several weeks and participating in some activities. However in Macedonia, there were too few livestock breeders and hunters per village. Therefore, we had to visit 33 villages to obtain an acceptable number of informants and participant observation was not possible given the limited time.

In both countries we used semi-structured in-depth interviews to explore local knowledge, perceptions and practices linked to the wolf. This method was chosen because 1) we wanted to obtain informants' categories with as little influence as possible from the ethnographer, and 2) we were focused on a particular topic, i.e. the human – wolf relationships. The use of an interview guide ensured that we addressed each subtopic with informants but we freely covered other subtopics when they took interesting directions (Huntington 2000). Sometimes we were not able to address some questions because of lack of time, misunderstanding, or a lack of will from the informant to answer the particular question.

Even if focused on local knowledge about wolf behaviour and ecology, the interview guide was containing questions concerning 1) daily activities, especially those linked with livestock breeding, and 2) local categorization of animals, space, and landscape in order to see if there is a distinction between nature and human, wild and domestic, and any place considered as wilderness, and where different animals are categorized inside this division of space. We were also particularly interested in reports of interactions with wolves in different contexts.

Interviews were conducted by the ethnographer (NL) accompanied by a native Kyrgyz translator in Kyrgyzstan and by native Albanian or Macedonian speakers in Macedonia, entirely recorded, and later transcribed and translated. While in Kyrgyzstan the interviews were translated with the help of the accompanying translator, in Macedonia we tended to use a different translator for the field and for the translation, thus allowing a cross-checking of the interviews.

A typical interview lasted for one hour or more and they were conducted in cafes, in private homes or directly in the field. In Kyrgyzstan, a total of 77 people were interviewed, mainly livestock breeders who were often hunters too, and also some folk medicine practitioners. In Macedonia, a total of 97 people were interviewed, including 57 livestock breeders, and 39 hunters. When historical resources were available, we also made some investigations into the development of livestock breeding and hunting activities and we also went through various statistical, government and NGO reports concerning livestock breeding in the two countries.

# RESULTS AND DISCUSSION

The first section of results largely summarises the changes that have occurred during post-communist transitions as described in public records and statistics, while the second part concerns the ways rural inhabitants, especially livestock breeders, perceive this change and the ways it has influenced their relationship with wolves.

## A Transition Process Affecting the Context of Human – wolf Relationships

### A General Collapse of Sheep Breeding Activities

Our investigations highlighted that the institutional and economic crisis following the collapse of the USSR and Yugoslavia had a strong impact on livestock breeding and hunting activities which were mainly dependent on the State before the collapse of USSR and Yugoslavia. Sheep breeding was the agricultural activity most affected by this crisis. Between 1992 and 2006, the number of sheep was halved in Macedonia and reduced to a third in Kyrgyzstan.

In Kyrgyzstan, the transition to a market economy was quite abrupt as in many former Soviet Republics in Central Asia (cf. Dear et al. 2012). Before the transition, all of the land was state property and most of the sheep (77%) were owned by the State and collective farms (Van Veen 1995). At the beginning of the transition process, many flocks were sold to pay collective farm debts in a context of high inflation, or sent to slaughter houses by unscrupulous chiefs or privatized on a patronage network basis (Jacquesson 2004). The number of sheep reached its minimum in 2004 and started to slowly increase from that year (FAOSTAT 2012). The situation was somewhat different in Macedonia, a place with a long tradition of pastoralism (Hadjigeorgiou 2011), as peasants in Yugoslavia had the choice to become private producers already in the 1950's (Boyd 1987). Therefore, even before transition, 90% of the sheep were privately owned even if State or collective farms (*AgroKombinat*) had large flocks ranging between 1,000 and 25,000 sheep (MAFWE 2003). The first reason for a post-communist decrease in sheep in Macedonia was the EU ban on the import of lamb meat from the country following the outbreak of foot-and-mouth disease in 1996 (Dimitrievski and Ericson 2010). Subsequently, the low prices of meat, milk, and wool combined with the high prices of fodder and concentrate feed, as well as the rising of costs of labour pushed the sheep breeders to reduce the size of their flocks (Dimitrievski and Ericson 2010).

## *Contrasting Trajectories of Rural Communities*

Although Kyrgyzstan and Macedonia are both undergoing changes in the dynamics of their rural populations, its nature and intensity are quite different in the two countries. In Yugoslavia, an economic and cultural policy of centralization as well as an accelerated industrialization starting in the 1960's led to an exodus from the countryside to towns and a concomitant decrease in the agricultural sector (Hadživukovi 1989). Shepherds also went to work in other countries, like in Italy, where their knowledge is appreciated (Pardini and Nori 2011). At the time of the Russian revolution, Kyrgyzstan was mainly occupied by nomadic livestock breeders and Russian colonists. After the restitution of colonists' land to the Kyrgyz people in 1921–1922 a process of collectivisation and the creation of kolkhozes was initiated from 1925 to 1932, which was accompanied by the settlement of the nomads (Jacquesson 2004). As the Kyrgyz economy remained focused on livestock breeding, there was no massive rural exodus like in Yugoslavia. Moreover, even if the proportion of rural population in Kyrgyzstan is slightly decreasing, the rural population growth rate remains positive until now while it started to be negative as early as the 1960's in Yugoslavia. As a consequence, the proportion of the population living in rural areas in Macedonia dropped from 76.6% in 1950 to 27.7% in 2010 (FAOSTAT2012) while in Kyrgyzstan it only decreased from 73.5% to 63.4% in the same period (FAOSTAT2012), implying that Kyrgyzstan remains a rural country while Macedonia became mainly urban from the 1970's. In Macedonia, there is a strong rural abandonment, a continuous decrease of the rural population (with a rural population growth rate of −2.2% during the 2005–2010 period, cf. FAOSTAT2012), and a strong decline in the rural way of life, with a feeling of lack of respect for rural people and livestock breeders.

## Changes in Hunting Pressure on Wolves

The post-communist upheaval in the USSR and Yugoslavia also had an impact on hunting activities. It is highly probable that this impact was particularly strong in Kyrgyzstan where hunting was partly an economic activity providing meat, fat, and fur. Each kolkhoz used to have its professional hunters, and shepherds were also equipped with rifles. Looking at the general picture in the USSR, it is obvious that wolf hunting was highly organized at the State level and subsidised (Bibikov 1973, 1980, Bibikov et al.1983). Wolves were considered as pest animals and trapped, captured in dens, poisoned and eventually hunted from planes or helicopters in open areas. The economic and logistical means supporting this intensive wolf hunting were no longer available after the collapse of the USSR. In Kyrgyzstan, the membership of the hunting association dramatically decreased from 25,900 in 1990 to 8,617 in 2002 as a consequence of Russian emigration from Kyrgyzstan. This decrease could be compensated by an increase in poaching from non-registered hunters, but it appears that the possession of small arms (legal or illegal) in the country is quite low (MacFarlane et al.2004). As a consequence, it is highly probable that the hunting pressure on wolves really decreased after the independence of Kyrgyzstan, as in other former Soviet countries (e.g. Belarus, cf. Sidorovich et al.2003).

On the other hand, the decrease in livestock numbers could have had an impact on wolf populations, even if the scarce data available on wolf diet in Kyrgyzstan doesn't show a high proportion of domestic animals (ca. 15% in the Central Tian Shan according to Vyrypajev and Vorobjev 1983). The level of poaching and its impact on wild ungulates and marmots (*Marmota baibacina*, *M. caudate*, *M.menzbieri*), which are the wolf's main wild prey in Kyrgyzstan, remains unknown. It certainly increased for highly valuable species like snow leopard (*Uncia uncia*) and probably also for argali sheep (*Ovis ammon*) and ibex (*Capra ibex*) which can provide meat and valuable trophies (Koshkarev 1994).

As for livestock breeding, the impact of the transition process on the organization of hunting was probably less dramatic in former Yugoslavia than in Kyrgyzstan. Macedonia is divided into 249 hunting grounds and apart from the State hunting grounds, an open competition was held in 2002 to award concessions to the highest bidders (Petkovski et al. 2003). Users of the hunting ground have to pay for the management plan as well as an annual fee of 20% of the estimated value of the game present within the hunting ground (Petkovski et al. 2003). According to Petkovski et al. 2003, hunters are largely dissatisfied since they have to pay for expensive management plans when at the same time the legal system doesn't ensure the punishment of poachers. Therefore, many users are not paying their membership fees and poaching is considerably higher than before independence as a result of the lack of an organised game warden service. A report from the Ministry of Environment (Ministry of Environment and Physical Planning 2003) also bemoans the fact that despite the existence of hunting management plans and a Public Enterprise for Game Wardens and Hunting Inspections, poaching remains at a high level. Concerning the wolf, the number of wolves reported as being killed has slowly increased since the 1960's and although there are no statistics for the transition period 1988–1992, it jumped from 200 wolves killed in 1987 to 460 in 1994, showing that hunting pressure on wolves didn't decrease, except maybe during the 1988–1992 period. Current harvests (2008–2010) have been between 108 and 188 wolves killed per year. Nowadays, wolves are classified as pests in both countries and there is a bounty for killed wolves. However, wolf hunting requires good hunting skills and time. In Macedonia, wolves are mainly killed by chance when hunting other animals while in Kyrgyzstan they are still actively hunted by knowledgeable hunters even though the bounty is not really a motivation since it is not always paid because of corruption.

## Redefining the Place of the Wolf in a Changing Socio-Economic and Environmental Context

Although both Kyrgyzstan and Macedonia went through radical transition processes, these had different influences on the relationships between livestock breeders and wolves. In Kyrgyzstan, the human population remains predominantly rural and livestock breeding remains one of the main activities in the country, and wolves are considered as one of the main threats for livestock breeders. In Macedonia, rural abandonment has been dramatic and livestock breeders are becoming increasingly rare. They feel isolated and ignored by the State and they have to face economic difficulties which are endangering this activity. We illustrate these contrasting development paths through a series of narratives that emerged from our fieldwork.

## Wolves as One of the Main Threats To the Capital of Kyrgyz Villagers

As we saw in the first part, Kyrgyzstan had to face a hard and dramatic transition process. Many people complained about the economic situation and all the advantages they lost with the collapse of Soviet Union, especially the social security brought by the State and the job security they had being employees of kolkhoz or sovkhoz (cf. Anderson and Pomfret 2000for economic analysis of Kyrgyz households during Transition Process).

However, despite these changes, livestock breeding remains one of the main activities in the country, accounting for 44% of the total agricultural output in 2004 (World Bank 2005). In addition, with the loss of social security and as a consequence of very low pensions and salaries, livestock became a vital form of capital and most villagers have a few sheep, one or two cows and sometimes a few horses. Cows are mainly milked for family use, and mares are milked for producing fermented mare's milk (*kymyz*), a highly valued drink in Kyrgyz society. Other animals are mainly kept as a capital which can be sold in case of important expenditures (school and university fees, hospitalization, etc.). Under these conditions, villagers are always trying to increase the size of their flock, since it means increasing their capital (cf. Liechti 2012too for the change in the meaning of livestock). As a consequence, when a wolf attacks livestock, it is not only threatening an ongoing commercial activity which produces income, but also villagers' capital. Therefore, even if taking into account the bias brought by the fact the surveys were focused on wolves, it appears that these animals are considered as one of the main threats to livestock breeding activities:

"In our area, the wolf is most important. The most competent, the most predatory, is the wolf. He is of no use, he is very harmful. He eats livestock if we don't put it in the barn" (Shepherd from Ak-Terek, KG: 05/2004)

The impact is enhanced because the general economic situation of the country is pushing people to poverty:

"They are harmful to people. It is a time of destitution and they eat the foal of one person, the sheep of another one, and that's it, it is very hard. People go to the market all the day to earn five som (ca. 0.12 USD)." (Shepherd from A a-Kajyndy, KG: 03/2004)

Therefore, when Kyrgyz shepherds recall the Soviet time, they point out that the wolf "problem" was better managed at that time: "At that time, hunters tried to reduce [the wolves] whereas now there is no help from the State for that" (Hunter from At-Bašy, KG: 11/2005). Now, "there is no one from the hunting inspectorate hunting wolves like before. There is no hunter hunting wolves with traps. There are no hunters like before. They come and they say they belong to the hunting inspectorate and that's it, they don't do anything" (Shepherd from Kara-Saz, KG: 07/2003). Not only do the official hunters not hunt wolves, but the shepherds can't have a rifle anymore and so can't scare or kill attacking wolves. During the Soviet period all shepherds had a rifle, although only official hunters could hunt wolves, now everybody can hunt the wolf but very few shepherds have a license for a rifle. Moreover, hunters with licenses don't always have enough money to buy bullets:

"There is no help from the State anymore. Hunting inspectors don't have rifles, they don't have bullets, people don't have means, they don't have bullets, and that's why [the wolves] are coming closer." (Shepherd from A a-Kajyndy, KG: 11/2005)

In addition to hunting activities, the overall presence of armed shepherds in the landscape was perceived as ensuring that wolves had to withdraw back into areas with less human activities. Indeed, during the soviet era the human presence in the landscape was more important. It was possible for livestock breeders to graze their flock in high valleys (called *syrt* in Kyrgyz) and stay there during the winter thanks to well-maintained infrastructure and provisioning with helicopters. As it is no longer possible, these high valleys are no longer occupied and generally, livestock breeders are going to summer pastures closer to villages because it is less costly (Jacquesson 2004, Crewett 2012). To some extent, it is now humans who have to withdraw back into areas close to their village and wolves which are now coming down close to villages:

"During the Kolkhoz, there were 52,000 sheep. They were on summer pastures, at the mountains' foothills, whereas now there is nothing, all the livestock is in the village and there is not much livestock, so of course [the wolves] come down..." (Shepherd from Terek-Suu, KG: 11/2005)

"Now, [the wolves] have adapted to these shrubs whereas before, near these shrubs, there were people, the Kolkhoz's livestock, and people were shooting with the rifle, whereas now [the wolves] are free, they are installed among these shrubs." (Shepherd from Žer-Uj, KG: 12/2005)

If Kyrgyz livestock breeders point to the impact that the transition process had on the wolves' population and behaviour and in turn the strong impacts wolves have on their activities, they nevertheless concede that there were many attacks during the Soviet time as well, but as it was Kolkhoz or Sovkhoz livestock, and not their own, it was therefore perceived as being less problematic:

"During the Soviet Era, as they were a lot of livestock at the Sovkhoz, at the Kolkhoz, we weren't feeling that much. [Wolves] were eating, but the shepherds were paying that, without informing people. And then we started to feel it after the fall of the Soviet, when they started attacking private livestock." (Shepherd from Ak-Terek, KG: 05/2004)

"Before, there used to be 30,000 to 100,000 sheep in the Kolkhoz, and wolves were a lot of nuisance. Now the wolf eats only one cow, only one foal from people, and it is difficult for them." (Shepherd from Tört-Kül, KG: 04/2004)

Thus, each attack on livestock has a more serious impact on their lives than before in terms of both personal economy and perception. In addition they have the feeling they can't effectively control wolves anymore:

"That's it. We don't manage to control them. At the time when I was shepherd, one could only see wolves with binoculars. Now they go very unhurriedly, they stop, they watch you. It is when you don't have rifle with you. Now there are no rifles and they know that." (Shepherd from A a-Kajyndy, KG: 03/2004)

As a consequence, wolves are now reported to approach villages and attack livestock in the immediate surroundings:

"Otherwise, now they are too close to the villages, they are around At-Bašy and they eat livestock. They cause many troubles to livestock." (Hunter from At-Bašy, KG: 11/2005)

Thus, it appears that rapid social changes in the context of the Kyrgyz transition to the market economy has had a clear impact on human – wolf relationships since it has made livestock breeders more vulnerable to wolf attacks and more prone to be highly affected by these attacks since each domestic animal is economically more valuable than before in this difficult period.

As a consequence, wolves are perceived as a main threat to livestock breeding and thus to most villagers who generally have a few livestock. Despite this situation, many Kyrgyz villagers still have a clear view that wolves shouldn't be eradicated since predators belong to nature and are regarded as having a sanitary role:

"It is good to have predators. They are nature's beauty so why eradicate them? It is good when they exist. It is good for children who come after and in some places, they have been eliminated, like tigers. Tigers were numerous close to the lake, in Tüp. Now there is no tiger anymore." (Retired hunter from Tört-Kül, KG: 04/2004)

"No, there is just a need to decrease the quantity [of wolves] because they eat sick animals and carrion that remains on the fields." (Shepherd/hunter from A a-Kajyndy, KG: 03/2004)

However, this view on the ecological role of the wolf was contested by some:

"Yes, [the wolves] have to be eradicated. All of them! What utility do they have? It is said they are cleaning the fields but to clean the fields we can take foxes, vultures and raven, they eat carrion. I never heard that wolves eat carrion. They only eat livestock. They eat livestock and make carrion, and the rest is eaten by vultures." (shepherd from Žer-Uj, KG: 04/2004)

And furthermore, the wolf impact on livestock is regarded as being hard to bear considering the economic difficulties:

"In these times, [wolves] are an animal we have to eliminate, if possible. Now people don't have many livestock like before." (Shepherd from Tört-Kül, KG: 04/2004)

"Right now we don't need wolves. They don't have any utility. [...] They eat foals, they eat livestock without sickness. It is not a sanitary animal. Wolves eat the cow, the calf from a poor man. See what losses it brings!" (Shepherd, Bel Tam 04/2004)

As a consequence, the wolf is even seen as threatening the Kyrgyz way of life:

"Outside, yaks are eaten by wolves. So we have to keep them in enclosures, like sheep, like horses. So you see neither horses, nor yaks, nor cows outside. So if we don't eliminate the wolves, what do we do with them? The State could give means. If not where are we going?" (shepherd from Korgondu-Bulak, KG: 12/2005)

## Wolves as an Additional Threat to a Highly Weakened Activity in Macedonia

The situation in Macedonia appears to be different to the Kyrgyz one. Despite the fact that livestock breeding remains a relatively important economic activity in the country, accounting for 24% of the total agricultural output in the period 1995–2007 (Dimitrievski and Ericson 2010), livestock breeders are not numerous in the country and the mean size of the owned flocks was generally bigger, at least in the Shara Mountains, than in Kyrgyzstan. Livestock breeding and especially sheep breeding appears to be a more commercial (rather than subsistence) activity in northwestern Macedonia, especially accounting for the fact that it is focused on cheese and lamb production in a highly seasonal manner, while in Kyrgyzstan there is no production of sheep cheese, and mutton and/or lamb meat is consumed year round.

In this situation, sheep are not the main capital for most Macedonian villagers, but rather the object of commercial activity for a few professional livestock breeders, at least in the investigated regions. Looking at the results of our investigations, it thus appeared that livestock breeders are mainly complaining about the difficulties inherent to economic activities linked with the transition process and its impact on livestock breeding.

Their first complaint is often about the lack of markets or access to markets. There is a general view that it was better before the collapse of Yugoslavia, when the State bought the cheese and the meat, and "public enterprises were also buying the wool" (shepherd from Pojarane, MK: 05/2008). Not only "were they coming to take the lambs on site" but "if one needed fodder, they could take from this firm" (shepherd from Dobri Dol, MK: 10/2007), and "the State was putting fertilizer on mountain pastures. We were paid to work on mountain pastures, even to mow in some places" (shepherd from Dobri Dol, MK: 11/2007). But "Now there is nothing, no activities" (idem). Not only does the State no longer ensure the market for sheep breeders' products, but also the domestic consumption of lamb and mutton in Macedonia has decreased. It went from 10.1kg in 1995 to 3.3kg per household in 2007, mainly because of the increased price as a consequence of decreasing sheep production and increased exports (Dimitrievski and Ericson 2010). The domestic consumption of sheep milk and cheese in the country also decreased from 15,643 to 11,291 tons in the same period (idem). Despite the export of lamb and mutton and the increased producer prices, sheep production doesn't cover the large expenses linked with extensive sheep breeding. It is especially important in the northwestern mountainous regions where livestock breeders are notably complaining about the lack of infrastructure to access their high pasture. In several cases the trail going to the summer pasture is only accessible using horses, making it impossible for them to go back home every day and/or to easily transport materials and to bring back milk or cheese.

According to livestock breeders, the market for lamb meat is not stable and exposes livestock breeders to economic risks when they don't manage to sell their lamb quickly. Nobody is interested in buying old lambs. In addition, the expenses are significant and the prices are growing. There is a need for labour to graze and milk sheep and also for additional food like hay and forage in winter in order to be able to produce lambs for the New Year, when their price is higher. Even if they generally manage to produce their own hay, they have to buy supplementary fodder, the price of which has strongly increased: "In previous years, we fed them with fodder like maize, barley, and compound feed, but now, the prices of fodder are quite high" (shepherd from Šipkovica, MK: 11/2007). Therefore, some livestock breeders can't feed them with fodder anymore and then have seen their production decrease, entering a vicious circle potentially driving them to bankruptcy. As a striking consequence of this catastrophic economic situation, most livestock breeders can't invest in or develop their activity, as this man told us:

"But if we would have more sheep, we would have more expenses! In this situation, it would be worse" (shepherd from Dobri Dol, MK: 10/2007)

The situation in Macedonia almost appears as the reverse of the Kyrgyz one. It is difficult for Macedonian livestock breeders to maintain their activity as an economically efficient one and increasing the size of their flock wouldn't help.

In this tight economic context, any additional cost is an additional weight that can't be tolerated. The wolf is considered as one of these additional costs and doesn't enjoy a positive image among hunters and livestock breeders. Indeed, interactions with wolves are mainly linked to livestock, and shepherds encounter wolves more frequently than hunters (Lescureux and Linnell 2010). As this shepherd explains: "Yes, as we are living in the summer pasture, we often see them, we are living with them!" (Shepherd from Lomincë, MK: 10/2007). For the western part of Macedonia, an interview survey among villagers conducted during 2006 reports 566 attacks on livestock from wolves (Keçi et al.2008). Not only are wolves considered as harmful for livestock, but this image is reinforced by the fact they can kill multiple sheep in each attack:

"… and the wolf too, if he takes a sheep, it doesn't mean anything, because he also has to live, but when he comes into the place where they are [enclosed], he doesn't take only one, he kills them all, all the ones he find… he cleans up. We had a case here when he killed 70 in one night!" (Hunter from Žurževice, MK: 10/2007)

Even if very few livestock breeders experience that type of surplus killing, the bad reputation of the wolf is widespread and gave rise to this idiom present in Albania and Macedonia claiming that "the wolf will kill 99 sheep and die at the hundredth" (See also Elsie 2001).

In addition to the direct costs linked with attacks on sheep, wolves have an indirect cost through the necessity for livestock breeders to keep livestock guarding dogs:

"[It is better] not having them [the wolves]! Because of the wolves, we feed seven to eight dogs. It is only because of them [the wolves] that we keep dogs. Without dogs we can't go on" (shepherd from Toplica, MK: 10/2007)

Indeed, even if only fed with maize flour or old bread mixed with whey, there is a cost to maintaining livestock guarding dogs (around 1 per dog per day), especially when they often have between 5 or more dogs per flock. Livestock breeders generally consider they would keep dogs even in the absence of wolves, for example to protect against theft, but they would then just keep one or two.

Despite wolves being harmful for livestock and costly for livestock breeders, they don't appear as the major threat for livestock breeding activities. As this sheep breeder remarked:

"No, problems with wild animals, it can be fixed. You know how to fix it. You need to pay more attention, you need more work, less sleep and you manage that way. It is the smallest problem." (Shepherd from Dobri Dol, MK: 10/2007)

However, the economic reasons are not the only ones pushing livestock breeders to have a negative perception of wolves.

## *Wolves as a Symbol of the Loss of Control over Nature in Macedonia*

Indeed, the transition process is not just affecting livestock breeders economically. The ongoing rural abandonment process and the crisis in livestock breeding are also affecting the landscape around the villages as well as the social structures in the villages and in the countryside. Therefore, the livestock breeders' perceptions of their social and natural environment are clearly affected and also their feelings concerning the future of their activity. Thus, for people who are traditionally livestock breeders and who have known the period when sheep breeding was flourishing, it is a traumatic experience to see this activity disappearing, since it is not only an economic activity but also a part of their traditions and their identity:

"Sheep are nowhere now. Investment is nowhere. We used to have the slaughterhouse in Gostivar and there we brought the wool, the cheese, the lambs [...]. Now there is nothing, neither slaughterhouse, nor anything else. We, for example, have had sheep from our ancestors' times and now there is nothing." (shepherd from Vrap ište, MK: 11/2007)

Livestock breeding is not the only activity disappearing from the villages. All the study villages from the Shara Mountains were surrounded by fields and orchards which are now mostly abandoned. As this livestock breeder said:

"Everything natural from which we were obtaining our bread in Kalište, it is abandoned now [...], I don't know how it is in France or in the West, I don't know there, but in this region, these people are disregarding the cultivated fields... the ground, the fields have been left abandoned. The government is not present here, why stay?" (idem)

As a consequence, there is a feeling that the surroundings of the villages are becoming wild:

"There are more juniper trees, more bushes, everything becomes wilder, even the pastures. And the fields became wilder. We have 640ha [of fields around the village], everything became pasture. We are using more fields than alpine pastures, because fields are also covered with junipers. Because we don't work in the fields, we don't work anything there and that's why." (shepherd from Vešala, MK: 11/2007)

Moreover, livestock breeders feel particularly isolated and unable to arouse interest from the government, the politicians, or society in general. Some of them are desperate about this situation: "Now, nobody is asking us, nobody is even looking at us" (shepherd from Pojarane, MK: 05/2008) and in general they feel there is a lack of recognition of the value of livestock breeding activities:

"Another important thing is that those who decide to do that job [shepherds] are unfortunately not well paid. They are not valued. [...] Here, shepherds are considered as one of the lower class, but it is a job like any other, and it deserves to be well paid." (shepherd from Gali nik, MK: 05/2008)

Thus, pastures are suffering shrub encroachment; fields are becoming pastures and the former system of order is being disrupted. The situation is very similar to what Höchtl et al. observed in the southwestern Alps:

"Many people felt wronged by politicians and are very unhappy about this situation. The decreasing usability and accessibility of the landscape leads to a loss of historical experience, cultural knowledge and local identity; in short, the land is increasingly losing its value as a "homeland"." (Höchtl et al.2005)

As a consequence, there is a complete lack of confidence in the future:

"We do this job because we don't have any other choice. If I want to sell the sheep, nobody would buy them, and I can't change. It is difficult even for the future... if things continue in that state as until now, I have no confidence about the future." (shepherd from Zajas, MK: 05/2008)

Thus, most livestock breeders don't want their children to follow their path: "I have children of 10 years old and I tell them to study, to go to school, and not to take that job [of livestock breeder]" (shepherd from Šipkovica, MK: 11/2007), which is not very surprising when they have lost interest and confidence in themselves and their job:

"Well, [this] job doesn't really please me, but I don't know what else to do, because we don't have enough profit, we pay the fodder, we pay the employees, we pay this, we pay that, and then there is no market for our product..." (shepherd from Vrap ište, MK: 11/2007)

Therefore, many livestock breeders have the feeling that their activity is coming to an end, and that in this situation, it is an unsustainable activity apart from the biggest and richest livestock owners having access to subsidies and to the market who can survive.

In this situation of rural areas becoming wilder and more and more hostile for human activities, wolves appear as a wild animal particularly difficult to control and symbolise the intrusion of the wild into the domestic. Indeed, these animals often come to take several sheep and their damage can be relatively important at an individual or regional level. In addition to livestock, wolves often kill hunting dogs and they are also blamed for damage to populations of game animals.

As a result of their perceived damage to wild and domestic animals, and their harmfulness to nature in general, wolves are described as unprofitable monsters (Lescureux and Linnell 2010). Even if wolf hunting is authorized and encouraged, wolves are not easy to hunt as this livestock breeder explains: "I agree to eliminate the wolves, but where to find them? They are masters (*ustah* in Turk), and the wolves are harmful." (Livestock breeder from Vrap iste, MK: 11/2007). The difficulties inherent to wolf hunting in steep, forested habitats hinders the ability of humans to react to wolf attacks and thus weakens the reciprocity of human – wolf relationships (cf. Lescureux and Linnell 2010).

If wolves are not the main threat, they are nonetheless an additional threat to a declining activity which is already burdened with unsustainable costs. Their repeated and fatal intrusions into the domestic space create the impression of an animal that is disrespectful of borders and norms (cf.Knight 2000). Therefore they reinforce the fuzziness of the wild – domestic border in an ongoing process of reforestation and shrub encroachment. Harmful for livestock breeders, the wolf also appears as a symbol of land degradation and rural abandonment. As a consequence, a large majority of the livestock breeders we interviewed were in favour of their complete eradication in the country, since they don't get any use of them and the place would be quieter without them:

"It is necessary to eliminate them all, to hunt them all, because we don't get any use of them, we don't get anything from them. They are only doing damages, they don't give anything. They can only destroy." (shepherd from Vešala, MK: 11/2007)

"But, if we could eradicate them [the wolves], good for us! Then we would be freer with sheep!" (shepherd from Toplica, MK: 10/2007)

# CONCLUSIONS

The transition process has clearly affected both livestock breeding and hunting practices and therefore has had an influence on the frames of human - wolf relationships in these two countries. Moreover, beyond the direct impact, the transition process has generated different socio-economic and environmental contexts in which the place of the wolf has changed when compared to the previous periods under communist control. It therefore appears that changes in livestock breeding and hunting activities brought about by the post-communist transitions have in turn had an impact on human – wolf relationships. In order to understand their impact on the human perceptions of wolves, we had to place all those changes in their specific social and economic context.

In Kyrgyzstan, livestock breeding is one of the main economic activities, and is well regarded. Flocks are a real capital for rural people. In their view of the world, inherited from their shamanistic origins, there are no strong borders between the human world and the animal world, and wolves are considered as intelligent, conscious and even useful animals, removing carrion and killing sick ungulates. The wolf is thus an alter ego engaged in reciprocal relationships with humans (Lescureux 2006, 2007). However, the lack of control over wolves which now occurs is viewed as a loss of reciprocity and a break-down in the balance of the human – wolf relationship, all the more so as livestock breeders are also often hunters. Finally, wolves are no longer perceived as a respectable enemy they have to regulate, but more and more as an invader, preventing the increase of a herder's capital, and even threatening the future of pastoralism and economy in the country:

The situation in Macedonia is different. Sheep are not the main capital, and due to the lack of outlets for cheese and lambs, livestock breeders tend to decrease the size of their flocks to maximize the profit, or at least minimise costs. Moreover, extensive herding of sheep is a marginal activity in the country, and livestock breeders have the feeling they are marginalized and not a concern of the state. While bears are viewed as a kind of alter-ego, wolves have a very different image (Lescureux and Linnell 2010, Lescureux et al. 2011). They are perceived as bloodthirsty and pest animals, destroying livestock, hunting dogs and game animals. In the difficult economic and social context, wolves are not perceived as the main threat to rural life, but are regarded as an additional threat, which symbolizes their loss of control over nature, which is mainly linked to rural abandonment, and therefore have to be eliminated, as wolves have absolutely no usefulness for them.

In both countries contemporary perceptions of wolves don't appear to be a result of the persistence of negative beliefs, but more as a dynamic phenomenon linked with the current socio-economic situations, and also linked with the perceived ability of wolves to change their behaviour. Not only were our informants' perceptions of the wolf changing through time but they also had the perception that wolves are changing along with the wolf - human relationship. Therefore, the interactive properties of large carnivores and their possible impact on livestock breeding activities have to be taken into account when evaluating the impact of political and economic transition on pastoralist activities.

The underlying cultural context is also very important however, notably concerning the way people perceive the respective place – and role – of humans and animals in the environment. This question is notably linked with perceived borders between wild and domestic, a border which appears to be very dynamic at the landscape level. Indeed, in both countries we observed that humans have had to withdraw back from parts of the landscape that they were previously utilising (high valleys in Kyrgyzstan and shrubby pastures in Macedonia). At the same time wolves are coming back into these areas and the combination of these two phenomena makes the wolf a symbol of land abandonment and – in the case of Macedonia – rural depopulation.

Therefore, from the point of view of wolf conservation, it is likewise important to view their status and perception by local people within 1) a wider social economic context which is highly dynamic over time (Walker 2005), but also 2) a wider ecological context, taking into account the dynamic process occurring at the landscape level.

It also appears that pastoralism is linked with hunting and wildlife management and takes place in a broader socio-economic and political context prone to rapid changes. Therefore the understanding of pastoralists' conflicts with wildlife requires placing these conflicts into a system of relationships including different actors, both humans and non-humans, who can potentially influence or be affected by these conflicts. This type of analyse requires multidisciplinary efforts in order to combine findings from ecology and the social sciences.

Finally, the challenge for pastoralists is to maintain their activity facing these rapid changes. Despite the often divergent point of view on the topic of large carnivores, pastoralists and wildlife conservationists often are natural allies (Niamir-Fuller et al. 2012), although it is a challenge for both of them to find a sustainable coexistence between pastoralist activities and large carnivore presence.

# AUTHORS' CONTRIBUTIONS

NL participated in the conception and design of the research, acquisition, analysis and interpretation of data for both countries. JL was the project leader for Macedonia and participated in conception and design of research in this country. Both NL and JL have been involved in drafting the manuscript and revising it and have given their final approval of the version to be published.

# ACKNOWLEDGEMENTS

The field work in KG was funded by the Muséum National d'Histoire Naturelle (Paris). The field work in Macedonia has been funded by successive grants from the Fyssen Foundation, the Research Council of Norway and by an Intra-European Fellowship granted to NL from the EU's Marie Curie scheme. We would like to thank Nuraly Turganbaev from Kirghiz national University for his help in the field in Kyrgyzstan, Mustafa Sabit, Gjorge Ivanov, Dime Melovski, and Aleksandar Stojanov from Macedonian Ecological Society for their help in the field in Macedonia. We are also grateful to the two anonymous reviewers for their corrections and comments, which greatly helped us to improve the manuscript.

# REFERENCES

1.   Alleau, J. 2010. Une histoire du loup à l'époque moderne. Méthodes, sources et perspectives. In Repenser le sauvage grâce au retour du loup. Les sciences humaines interpellées, ed. Moriceau J-M, Madeline P, 23–39. Caen: Presses universitaires de Caen & MRSH.

2.   Anderson, K, and R Pomfret. 2000. Living standards during transition to a market economy: the Kyrgyz Republic in 1993 and 1996. Journal of Comparative Economics 28: 502–523.

3. Bangs, EE, SH Fritts, JA Fontaine, DW Smith, KM Murphy, CM Mack, and CC Niemeyer. 1998. Status of gray wolf restoration in Montana, Idaho, and Wyoming. Wildlife Society Bulletin 26: 785–798.

4. Bibikov, DI. 1973. The Wolf in the USSR. In First Working Meeting of Wolf Specialists and of the First International Conference on Conservation of the Wolf, ed. Pimlott HD, Morges, Switzerland: International Union for Conservation of Nature and Natural Resources.

5. Bibikov, DI. 1980. Wolves in the USSR. Natural History 89: 58–63.

6. Bibikov, DI, NG Ovsyannikov, and A Filimonov. 1983. The status and management of the wolf population in the USSR. Acta Zoologica Fennica 174: 269–271.

7. Bobbé, S. 1993. Hors statut, point de salut. Ours et loups en Espagne. Etudes Rurales 129–130: 59–72.

8. Bobbé, S. 1993. Ours, loup, chien errant en Espagne. Des couples dans le bestiaire. In Des bêtes et des hommes. Le rapport à l'animal, un jeu sur la distance, ed. Lizet B, Giordani GR, 211–226. Paris: édition du comité des travaux historiques et scientifiques.

9. Bobbé, S. 2003. Polémique autour d'un projet de zonage, appliqué à la gestion des loups dans l'arc alpin français. Espaces et sociétés 110–111: 111–128.

10. Boitani, L. 1992. Wolf research and conservation in Italy. Biological Conservation 61: 125–132.

11. Boitani, L. 1995. Ecological and cultural diversities in the evolution of wolf-human relationships. In Ecology and conservation of wolves in a changing world, ed. Carbyn LN, Fritts SH, Seip DR, 3–11. Edmonton, Alberta: Canadian Circumpolar Institute.

12. Boitani, L. 2003. Wolf conservation and recovery. In Wolves: behavior, ecology, and conservation, ed. Mech LD, Boitani L, 317–344. Chicago: The University of Chicago Press.

13. Boyd, ML. 1987. The performance of private and cooperative socialist organization: postwar Yugoslav agriculture. The Review of Economics and Statistics 69: 205–214

14. Breitenmoser, U. 1998. Large predators in the Alps: the fall and rise of man's competitors. Biological Conservation 83: 279–289.

15. Brox, O. 2000. Schismogenesis in the wilderness: the reintroduction of predators in Norwegian forests. Ethnos 65: 387–404.

16. Brunois, F. 2005. Man or animal: who copies who? Interspecific empathy and imitation among the Kasua of New Guinea. In Animal Names, ed. Minelli A, Ortalli G, Sanga G, 369–381. Venezia: Istituto Veneto di Scienze Lettere ed Arti.

17. Brunois, F. 2005. Pour une approche interactive des savoirs locaux : l'ethno-éthologie. Journal de la Société des Océanistes 120–121: 31–40.

18. Ciucci, P, L Boitani, F Francisci, and G Andreoli. 1997. Home range, activity and movements of a wolf pack in Central Italy. Journal of Zoology, London 243: 803–819.

19. Clark, TW, AP Curlee, and RP Reading. 1996. Crafting effective solution to the large carnivore conservation problem. Conservation Biology 10: 940–948.

20. Clark, TW, PC Paquet, and AP Curlee. 1996. Special section: large carnivore conservation in the Rocky Mountains of the United States and Canada. Introduction. Conservation Biology 10: 936–939.

21. Coleman, JT. 2004. Vicious: wolves and men in America. New Haven: Yale University Press.

22. Comincini, M, A Oriani, C Morbioli, R Castiglioni, and A Martinoli. 2002. L'uomo e la "bestia antropofaga". Storia del lupo nell'Italia settentrionale dal XV al XIX secolo. Milan: Unicopli

23. Crewett, W. 2012. Improving the Sustainability of Pasture Use in Kyrgyzstan. Mountain Research and Development 32: 267–274.

24. Dear, C, H Weyerhaeuser, H Hurni, SW von Dach, and A Zimmermann. 2012. Special issue: Central Asian mountain societies in transition. Mountain Research and Development 32: 265–266.

25. Descola, P. 2005. Par-delà nature et culture. Gallimard, Paris.

26. Descola, P, and G Pálsson. 1996. Nature and Society. Anthropological perspectives. London: Routledge.

27. Dimitrievski, D, and T Ericson. 2010. Sector Study - Macedonian Agriculture in the period 1995–2007. University Ss Cyril and Methodius & Swedish University of Agricultural Sciences.

28. Doré, A. 2010. L'histoire dans les méandres des publics : quand les "méchants loups" ressurgissent du passé. In Repenser le sauvage grâce au retour du loup. Les sciences humaines interpellées, ed. Moriceau J-M, Madeline P, 75–89. Caen: Presses Universitaires de Caen.

29. 2001. A dictionary of Albanian religion, mythology, and folk culture,Elsie R New York: New York University Press.

30. Falcucci, A, L Maiorano, and L Boitani. 2007. Changes in land-use/land-cover patterns in Italy and their implication for biodiversity conservation. Landscape Ecology 22: 617–631.

31. FAOSTAT. 2012. Food and Agriculture Organization. http://faostat3.fao.org/home/index.html#HOME *webcite*. Accessed 01 June 2012

32. Fritts, SH, EE Bangs, JA Fontaine, MR Johnson, MK Phillips, ED Koch, and JR Gunson. 1997. Planning and implementing a reintroduction of wolves to Yellowstone National Park and central Idaho. Restoration Ecology 5: 7–27.

33. Fritts, SH, RO Stephenson, RD Hayes, and L Boitani. 2003. Wolves and humans. In Wolves: behavior, ecology, and conservation, ed. Mech LD, Boitani L, 289–316. Chicago: The University of Chicago Press.

34. Hadjigeorgiou, I. 2011. Past, present and future of pastoralism in Greece. Pastoralism: Research, Policy and Practice 1: 24.10.1186/2041-7136-1-24 BioMed Central

35. Hadživukovi , S. 1989. Population growth and economic development: a case study of Yugoslavia. Journal of Population Economics 2: 225–234.

36. Haraway, D. 2003. The Companion Species Manifesto. Dogs, People, and Significant Otherness. Chicago: Prickly Paradigm Press.

37. Höchtl, F, S Lehringer, and W Konold. 2005. "Wilderness": what it means when it becomes reality - a case study from the southwestern Alps. Landscape and Urban Planning 70: 85–95.

38. Huntington, HP. 2000. Using traditional ecological knowledge in science: methods and applications. Ecological Applications 10: 1270–1274.

39. Ingold, T. 1996. Hunting and gathering as ways of perceiving the environment. In Redefining nature: ecology, culture and domestication, ed. Ellen R, Fukui K, 117–154. Oxford: Berg.

40. Ingold, T. 2000. From trust to domination. An alternative history of human-animal relations. In The perception of environment. Essays on livelihood, dwelling and skill, ed. Ingold T, 61–76. London and New York: Routledge.

41. Ivanov, G, A Stojanov, D Melovski, V Avukatov, E Keçi, A Trajçe, S Shumka, G Schwaderer, A Spangenberg, JDC Linnell, M von Arx, and U Breitenmoser. 2008. Conservation status of the critically endangered balkan lynx in Albania and Macedonia. Proceedings of the III congress of ecologists of the Republic of Macedonia with international participation. 249–256. Skopje (Macedonia): Macedonian Ecological Society.

42. Jacquesson, S. 2004. Au coeur du Tian Chan : histoire et devenir de la transhumance au Kirghizstan. Cahiers d'Asie Centrale 11 (12): 203–244.

43. Kaczensky, P. 1999. Large carnivore predation on livestock in Europe. Ursus 11: 59–72.

44. Keçi, E, A Trajçe, K Mersini, F Bego, G Ivanov, D Melovski, A Stojanov, U Breitenmoser, M VonArx, G Schwaderer, A Spangenberg, and JDC Linnell. 2008. Conflicts between lynx, other large carnivores, and humans in Macedonia and Albania. Proceedings of the III congress of ecologists of the Republic of Macedonia with international participation (06–09.10.2007). 257–264. Skopje: MES.

45. Kellert, SR, M Black, C Reid Rush, and AJ Bath. 1996. Human culture and large carnivore conservation in North America. Conservation Biology 10: 977–990.

46. Knight, J. 2000. Introduction. In Natural Enemies. People-wildlife conflicts in anthropological perspective, ed. Knight J, 1–36. London: Routledge.

47. Koshkarev, E. 1994. Poaching in Kyrgyzstan. Cat News, newsletter of the IUCN Cat Specialist Group 21.

48. Latour, B. 1996. On actor-network theory. A few clarifications plus more than a few complications. Soziale Welt 47: 369–381.

49. Lescureux, N. 2006. Towards the necessity of a new interactive approach integrating ethnology, ecology and ethology in the study of the relationship between Kirghiz stockbreeders and wolves. Social Science Information 45: 463–478.

50. Lescureux, N. 2007. Maintenir la réciprocité pour mieux coexister. Ethnographie du récit kirghiz des relations dynamiques entre les hommes et les loups. Paris: Muséum National d'Histoire Naturelle.

51. Lescureux, N, and JDC Linnell. 2010. Knowledge and perceptions of Macedonian hunters and herders: the influence of species specific ecology of bears, wolves, and lynx. Human Ecology 38: 389–399

52. Lescureux, N, JDC Linnell, D Melovski, A Stojanov, G Ivanov, and V Avukatov. 2011. The king of the forest. Local knowledge about European brown bears (Ursus arctos) and implications for its conservation in contemporary Western Macedonia. Conservation and Society 9: 189–201.

53. Liechti, K. 2012. The meanings of pasture in resource degradation negotiations: evidence from post-socialist rural Kyrgyzstan. Mountain Research and Development 32: 304–312.

54. Linnell, JDC, JE Swenson, and R Andersen. 2001. Predators and people: conservation of large carnivores is possible at high human densities if management policy is favourable. Animal Conservation 4: 345–349

55. Linnell, JDC, EJ Solberg, SM Brainerd, O Liberg, H Sand, P Wabakken, and I Kojola. 2003. Is the fear of wolves justified? a fennoscandian perspective. Acta Zoologica Lituanica 13: 34–40.

56. Linnell, JDC, U Breitenmoser, C Breitenmoser-Würsten, J Odden, and M von Arx. 2009. Recovery of Eurasian lynx in Europe: What part has reintroduction played? In Reintroduction of top-order predators, ed. Hayward M, Sommers M, 72–91. Oxford: Blackwell Publishing.

57.  Løe, J, and E Røskaft. 2004. Large carnivores and human safety: a review. Ambio 33: 283–288.

58.  Lohr, C, WB Ballard, and AJ Bath. 1996. Attitudes toward gray wolf reintroductions to New Brunswick. Wildlife Society Bulletin 24: 414–420.

59.  Lopez, BH. 1978. of wolves and men. New York: Charles Scribner's.

60.  MacFarlane, N, S Torjesen, and C Wille. 2004. An anomaly in Central Asia? Small Arms in Kyrgyzstan. In Small arms survey 2004 - rights at risk, ed. Batchelor P, Krause K, 309–324. New York: Oxford University Press.

61.  MAFWE. 2003. Country Report on the State of the Animal Genetic Resources in Republic of Macedonia. Republic of Macedonia Ministry of Agriculture, Forestry and Water Economy, .

62.  Mauz, I. 2005. Gens, cornes et crocs. Paris, Grenoble: Inra-Quae.

63.  Mech, LD. 1970. The wolf the ecology and behavior of an endangered species. Minneapolis: University of Minnesota Press.

64.  Mech, LD. 1995. The challenge and opportunity of recovering wolf populations. Conservation Biology 9: 270–278.

65.  Mertens, A, and C Promberger. 2001. Economic aspects of large carnivore-livestock conflicts in Romania. Ursus 12: 173–180.

66.  Ministry of Environment and Physical Planning. 2003. Biodiversity Strategy and Action Plan for the Republic of Macedonia.

67.  Moore, RS. 1994. Metaphors of encroachment: hunting for wolves on a Central Greek Mountain. Anthropological Quarterly 67: 81–88.

68.  Moriceau, J-M. 2007. Histoire du méchant loup. 3 000 attaques sur l'homme en France. Paris: Fayard.

69.  Niamir-Fuller, M, C Kerven, R Reid, and E Milner-Gulland. 2012. Co-existence of wildlife and pastoralism on extensive rangelands: competition or compatibility? Pastoralism: Research Policy and Practice 2012: 8

70.  Pardini, A, and M Nori. 2011. Agro-silvo-pastoral systems in Italy: integration and diversification. Pastoralism: Research, Policy and Practice 1: 26.

71. Petkovski, S, D Smith, T Petkovski, and V Sidorovska. 2003. Study on hunting activities in Macedonia: past, present and future. Society for the Investigation and Conservation of Biodiversity and the Sustainable Development of Natural Ecosystems (BIOECO).

72. Rajpurohit, KS. 1999. Child lifting: wolves in Hazaribagh, India. Ambio 28: 162–166.

73. Røskaft, E, T Bjerke, BP Kaltenborn, and JDC Linnell. 2003. Patterns of self reported fear towards large carnivores among the Norwegian public. Evolution and Human Behaviour 24: 184–198.

74. Røskaft, E, B Handel, T Bjerke, and BP Kaltenborn. 2007. Human attitudes towards large carnivores in Norway. Wildlife Biology 13: 172–185.

75. Salvatori, V, and JDC Linnell. 2005. Report on the conservation status and threats for wolf (Canis lupus) in Europe. Conseil de l'Europe - Council of Europe. Report T-PVS/Inf (2005) 16

76. 1998. Bears. Status Survey and Conservation Action Plan, Servheen C, Herrero S, Peyton B Switzerland & Cambridge, UK: IUCN/SSC Bear and Polar Bear Specialist Group, Gland.

77. Sidorovich, VE, LL Tikhomirova, and B J drzejewska. 2003. Wolf Canis lupus numbers, diet and damage to livestock in relation to hunting and ungulate abundance in northeastern Belarus during 1990–2000. Wildlife Biology 9: 103–111.

78. Skogen, K, I Mauz, and O Krange. 2008. Cry wolf!: narratives of wolf recovery in France and Norway. Rural Sociology 73: 105–133.

79. Theuerkauf, J. 2003. Impact of man on wolf behaviour in the Bialowieza Forest. Poland: Technischen Universität München, Munich.

80. Theuerkauf, J. 2009. What drives wolves: fear or hunger? humans, diet, climate and wolf activity patterns. Ethology 115: 649–657.

81. Theuerkauf, J, W Jedrzejewski, K Schmidt, and R Gula. 2003. Spatiotemporal segregation of wolves from humans in the Bialowieza forest (Poland). Journal of Wildlife Management 67: 706–716

82. Theuerkauf, J, W Jedrzejewski, K Schmidt, H Okarma, I Ruczynski, S Sniezko, and R Gula. 2003. Daily patterns and duration of wolf activity in the Bialowieza forest, Poland. Journal of Mammalogy 84: 243–253.

83. Treves, A, and KU Karanth. 2003. Human-carnivore conflict and perspectives on carnivore management worldwide. Conservation Biology 17: 1491–1499.

84. Van Veen, TWS. 1995. The Kyrgyz sheep herders at a crossroad. Pastoral Development Network Series 38: 1–14.

85. Vilà, C, V Urios, and J Castroviejo. 1995. Observations on the daily activity patterns in the Iberian Wolf. In Ecology and Conservation of Wolves in a Changing World, ed. Carbyn LN, Fritts SH, Seip DR, 335–340. Alberta: Canadian Circumpolar Institute.

86. Vyrypajev, VA, and GG Vorobjev. 1983. Volk v Kirgizii. Ilim, Frunze.

87. Walker, BL. 2005. The lost wolves of Japan. Seattle: University of Washington Press.

88. World Bank. 2005. Kyrgyz Republic. Livestock Sector Review: Embracing the New Challenges. World Bank.

# Citations

# CHAPTER 1

Ronald Glasberg, Michael Hartmann, Michael Draheim, Gerrit Tamm, and Franz Hessel, "Risks and Crises for Healthcare Providers: The Impact of Cloud Computing," The Scientific World Journal, vol. 2014, Article ID 524659, 7 pages, 2014. doi:10.1155/2014/524659.

# CHAPTER 2

Mark Errington and Stephen J Childe, A Business Process Model of Inspection in Remanufacturing, doi: 10.1186/2210-4690-3-7.

# CHAPTER 3

Alfredo Mela, Urban public space between fragmentation, control and conflict, doi: 10.1186/s40410-014-0015-0.

# CHAPTER 4

Kewen Wu, Zeinab Noorian, Julita Vassileva, and Ifeoma Adaji, How buyers perceive the credibility of advisors in online marketplace: review balance, review count and misattribution, doi:10.1186/s40493-015-0013-5.

# CHAPTER 5

Klaus Schuch, George Bonas, and Jörn Sonnenburg, Enhancing Science and Technology Cooperation between the EU and Eastern Europe as Well as Central Asia: A Critical Reflection on the White Paper from a S&T Policy Perspective, doi:10.1186/2192-5372-1-3.

# CHAPTER 6

Pengfei Ni, The Goal, Path, and Policy Responses of China's New Urbanization, doi: 10.1186/2196-5633-1-2.

# CHAPTER 7

Farzad Pour Rahimian, Tomasz Arciszewski, and Jack Steven Goulding, Successful Education for AEC Professionals: Case Study of Applying Immersive Game-Like Virtual Reality Interfaces, doi:10.1186/2213-7459-2-4.

# CHAPTER 8

Nicolas Lescureux and D John C Linnell, the Effect of Rapid Social Changes during Post-Communist Transition on Perceptions of the Human - wolf Relationships in Macedonia and Kyrgyzstan, doi: 10.1186/2041-7136-3-4.

# Index